Homo Lapsus

Sin, Evolution, *and the* God Who Is Love

By
Niamh M. Middleton

Deep River Books

Homo Lapsus: Sin, Evolution, and the God Who Is Love

Published by Deep River Books
Sisters, Oregon
www.deepriverbooks.com

The Scripture texts used in this work are taken from the New Revised Standard Version Bible, Catholic Edition (Sacramento CA: Harper One, 2010).

ISBN – 13: 9781632694881
Library of Congress: 2018961159

Printed in the USA

2018—First Edition

27 26 25 24 23 22 21 20 19 18 10 9 8 7 6 5 4 3 2 1

Cover design by Joe Bailen, Contajus Designs
Index compiled by James Helling

For Gerry

Contents

Preface

As a long-time fan of science fiction, I have always been intrigued by its power to generate the mythologies and philosophies associated with the greatest achievements of science. Indeed, at the 2016 convention in New York to celebrate the fiftieth anniversary of my favorite science fiction television series, *Star Trek*, I was struck by the fact that some leading actors of the original series have, in their own ways, become philosophers of science or moral philosophers. Even more than that, as a moral theologian and educator whose research interests are focused on the impact of science—biology in particular—on our understanding of human behavior, it was a revelation to find myself among a group of people made up of actors, writers, and astronauts who see the world through the same moral lens as I do. The field of the academic humanities in which I teach can be a difficult environment for someone who perceives and analyses human behavior through an evolutionary lens. It was particularly gratifying for me therefore to hear the actor William Shatner, who played the leading role of Captain Kirk in the original series, show himself to be at the cutting edge of contemporary moral philosophy in his grasp of the significance of evolution for our understanding of morality and in his highlighting of the need for educational institutions to take note of it. He went so far as to suggest that schools should teach classes in empathy, a quality which is known to be part of our evolutionary heritage but which, as he pointed out, is easily lost in the daily battle to survive in the highly competitive environments we have created for ourselves.

The split between science and the humanities (which includes the social sciences) over the former's evolutionary perspective on morality and the latter's corresponding neglect of it is, I believe, a serious obstacle to the progress of our species. The conflict is partly due, in my opinion, to rivalry over the rapid rise of scientism—the view that science is the highest form of knowledge—which began to gain momentum in the mid-twentieth century and continues apace. There is also the reluctance of the academic humanities to accept the possibility that there may be a definable human nature and that this nature may indeed be determined to a greater or lesser degree. The refusal to engage with science on the question of human behavior is dangerous, and not just because science now leads the field in its understanding of what makes us tick as human beings; there is also the fact that it may soon have the ability to engineer and modify human nature. In the field of reproductive technology, for example, science is already racing ahead of ethics at a frightening speed.

It is my profound conviction that the reconciliation of science and the humanities on the question of human nature and how it should be defined is the most important task now facing both fields. A significant breakthrough came recently in this regard from the Israeli author and history professor Yuval Noah Harari in his two books, *Sapiens* and *Homo Deus*. *Sapiens* in particular is groundbreaking not only because it is a best seller in popular science by an academic who works in the humanities but also because it challenges the conventional understanding of history to incorporate our prehistory (termed biology by Harari) hence creating a new narrative of humanity. Harari argues that we cannot fully understand ourselves or plan for our future progress as a species without also understanding how we evolved. Although written from an atheistic perspective, *Sapiens*' insights into the evolution of both morality and religion are highly pertinent to the religious domain; it has also provided a sound basis for the long-awaited synthesis between science and the humanities. Where this book parts company with Harari's is in its assertion that evolutionary biology corroborates foundational Christian doctrines and that while our flourishing as a species requires synthesis between science and the humanities in general, religion is the vital bridge between the two disciplines. I am in agreement with Harari's assertion that the wall separating biology departments in universities from the departments of law and politics must come down.[1] I am also concerned, however, that without input from the religious domain the removal of that barrier will actually be detrimental to human progress. This is a theme that is explored throughout the book in its analysis of the growing influence of science and scientism in public discourse.

Writing this book has been a long, challenging, and rewarding intellectual journey; in embarking on it I did not have to "boldly go where no (wo)man has gone before" regarding intellectual territory; many philosophers and theologians have produced excellent work in the interdisciplinary field of religion and science. What is new about my contribution to the task is that I have considered it from an angle that hasn't been tried before, enabling me to reach some surprising conclusions that

---

[1] Yuval Noah Harari, *Sapiens: A Brief History of Human*kind (London, Vintage Books, 2014), p. 263.

are nonetheless firmly rooted in the work of leading writers on the subject. It is my dearest hope that this addition to their efforts will be considered a worthy one.

INTRODUCTION

When the French Jesuit paleontologist and theologian Teilhard de Chardin wrote *The Phenomenon of Man* in the early twentieth century, he took a first, giant step in the reunification of religion and science. These two disciplines, which up until the early modern era had constituted a unified whole as fields of knowledge, gradually began to separate from one another as the amazing practical achievements of science took center stage in the public consciousness.

The first major challenge to their hitherto harmonious relationship was caused by Galileo's championing of heliocentric theory in the seventeenth century. Galileo, along with other geniuses of the early modern period such as Kepler and Newton, not only brought to the fore a new type of knowledge but challenged biblical creation stories by relegating the earth and its human inhabitants to insignificance in a mysterious, infinite universe. It was of course Copernicus, a Polish ecclesiastic, who had first identified the sun rather than the earth as the focal point of the solar system. During his productive years, the church was far more liberal than it became after the Reformation had taken hold; Copernicus actually dedicated his great work, *De Revolutionibus Orbium Coelestium*, to the pope of the time. Galileo however, defender and promoter of Copernicus's work, was tried over a century later for heresy by the revived inquisition, forced to recant under threat of torture, and sentenced to spend the final three decades of his life under house arrest. The rise of science was unstoppable, nonetheless; the combination of religious disillusionment caused by the wars of the Reformation era and a burgeoning confidence in human powers brought on by exciting scientific discoveries finally exploded in the eighteenth century into the intellectual movement known as the Enlightenment, a movement of such powerful impact that it led to the American and French revolutions. It was also a period of extreme optimism concerning human capabilities, during which the intellectual elite of Europe became convinced that humanity could solve its problems without recourse to a putative deity. Religious belief was gradually becoming optional, and the battle between religion and science for control over knowledge and definitions of "truth" was well and truly underway.

The second, perhaps even more seismic challenge came with the publication in 1859 of Charles Darwin's *On the Origin of Species*. Although evolution as a concept had become current and popular in the eighteenth century, it was Darwin who provided the mass of evidence that led to the acceptance by science of the evolutionary hypothesis that all life on Earth shares a common descent from a universal common ancestor. Darwin's evolutionary perspective, however, was radically different to that of the eighteenth-century intelligentsia whose view of evolution reflected Enlightenment optimism and culminated in the work of pioneering French biologist Jean Baptiste Lamarck.[2] The latter believed that evolution was pro-

---

[2] Although Lamarck published his *Philosophie Zoologique* in 1809, it was a reflection of the wisdom of the time accepted by many natural historians of the eighteenth century, who included Darwin's grandfather Erasmus Darwin.

gressive, with an inner drive toward perfection implanted in organisms by God, known as *orthogenesis*. This inner drive was propelled, according to Lamarck, by the passing on to offspring of favorable adaptive characteristics acquired by organisms during their lifetimes. The example that is most often given to illustrate Lamarckian evolution is that of a giraffe which, through constantly stretching its neck to reach the highest bananas in the tree, develops a longer neck. The extra length is then passed on to offspring, who increase the length a bit more, and so on. The inheritance of acquired characteristics as envisaged by Lamarck would imply a very speedy form of evolution. In the case of humans, an analogous example would be that moral characteristics such as courage and resilience, developed over the course of an individual's life in response to environmental exigencies, could be passed on to that individual's children. This would result in the rapid evolution of morality in our species.

Darwin however, along with his codiscoverer Alfred Russel Wallace, proposed that evolution is driven by a mechanism called *natural selection*. The principle of natural selection itself was developed by Darwin from his observations of breeders' use of artificial selection to improve their domestic animal stock. Then, having read T.R. Malthus's *Principle of Population*, Darwin saw that he could use Malthus's concept of the "struggle for existence" to explain the force that propels natural selection. Malthus's theory of competition within human populations over scarce resources was applied by Darwin in a general biological context. It provided him with a premise from which to argue to a universal struggle for existence. "Owing to this struggle," wrote Darwin,

> Any variation, however slight and from whatever cause proceeding, if it be in any degree profitable to an individual of any species, in its infinitely complex relations to other organic beings and to external nature, will tend to the preservation of that individual, and will generally be inherited by its offspring. The offspring also will thus have a better chance of surviving, for of the many individuals of any species which are periodically born, but a small number can survive.[3]

Those organisms (plant, animal, and human) that are the "fittest"—meaning the best able to adapt to their environments—will be "selected" by nature to survive and reproduce, until they predominate in and eventually transform a species. The term *adaptation* is a very important one in evolutionary biology. Adaptations are traits such as the structure of an organism, its physiology, or behavior that promote its survival and reproductive interests. Such favorable characteristics are not, however, acquired during the span of a single life but incrementally, by ancestral members of species, over vast evolutionary timespans. In contrast to the Lamarckian view of evolution, evolution by natural selection implies that life evolved slowly and gradually. Another implication is that random and chance events play a significant role in evolutionary change. Random mutations in genes can increase or decrease fitness depending on the environmental situation. Environmental challenges can occur due to chance events such as climate change or the

[3] Charles Darwin, *On The Origin of Species* (Harmondsworth: Penguin, 1968), p. 115.

isolation of small groups that are cut off from a larger population due to natural disasters or geographical features such as mountains. Another type of chance event can cause mass extinctions. When dinosaurs went extinct, probably due to the impact of a comet, mammals were able to evolve to be larger and more diverse, giving rise eventually to our own species. The implications of the Darwinian perspective on evolution were therefore stark: there is no overall purpose or meaning to creation, and no upward moral progress. Evolution is driven by competition between organisms, but this competition takes place within environments that are subject to chance and random events that favor some organisms over others.

Being, Knowing, and God

If science had previously challenged Christianity's monopoly on truth, however, it had not yet seriously disturbed the Christian worldview of the majority who chose to remain Christian. Indeed, during the Enlightenment period, British Anglicanism in particular developed a view of nature that was at once scientific and religious; nature, it was claimed, revealed God as both artist and scientist. The laws of nature that were being deciphered by science did not so much reflect the talents of scientists as point toward hitherto unappreciated powers of the deity. Nor had the concept of evolution yet endangered the synthesis that still existed between religion and biology concerning origins, which largely rested on mutual belief in either direct creation by God or Lamarckian orthogenesis. With the publication of *Origin* and, some years later, *The Descent of Man*, the Darwinian version of evolution quickly became popular, however. As a result, the underlying implication of an amoral universe to which a god or gods would be irrelevant even if they existed gradually began to permeate the public mind. Besides the threat posed by Darwinism to the Genesis creation accounts and to biblical interpretation generally, the controversy stirred up by *Origin* brought to the fore and crystallized another major issue at stake in the new battlefront between science and religion. The right to explain origins—and especially human origins—had traditionally belonged to the church. Now, certain brash practitioners of this (relatively) young discipline of science were appropriating that right for themselves and, by implication, strengthening the viewpoint that scientific knowledge has greater validity and claim to "truth" than the type of knowledge purveyed by the religious domain. The landmark Oxford debate that took place in June 1860 between Bishop Samuel Wilberforce and Thomas Henry Huxley, the famous young science writer who was known as "Darwin's bulldog" due to his appetite for battle, brought matters to a head. The repartee between the bishop and the scientist delighted the audience, and the contest quickly assumed legendary proportions; it has since gone down in scientific history as a turning point in the emergence of the discipline as an autonomous field. There is no doubt that Huxley achieved his goal, which was to establish science as an independent discipline with its own innate authority and the right to investigate natural phenomena without any reference to theological doctrines. By the mid-twentieth century, however, the balance of power between the two disciplines had begun to be reversed: contemporary public and political discourse is now increasingly influenced by scientism, a viewpoint that holds science to be the only valid form of knowledge. In tandem with scientism goes the view that human beings are totally material creatures, since science can now provide an explanation for the evolution of human intelligence and morality. The doctrine of evolution appears, in fact, to have torn to shreds both the ontological and the epistemological consensus that informed much of philosophical as well as theological discourse on human nature up until the early twentieth century. With regard to ontology (philosophy of being), humans were held to be a combination of matter and spirit. With regard to epistemology (philosophy of knowledge) therefore, human intellectual and moral abilities were considered to be in large part a product of the soul or spiritual dimension of the human being. Epistemologically speaking, revelation and *a priori* knowledge (i.e., knowledge that is self-evident and not dependent on sense experience) were generally thought to be the highest forms of knowledge.

It is no surprise that biology was the scientific discipline responsible for the decisive sundering of religion and science as fields of knowledge: biology has always overlapped more than other scientific disciplines with religion because one of its major concerns is the human species and how it relates to other forms of earthly life; biological definitions of humanity frequently clash with the spiritual claims made by the religious domain. Conflict between science and religion in the West tends to center therefore on the two related issues raised by evolutionary theory that pose the greatest challenge to its predominant religion, Christianity: these are creation and behavior. One of my aims in writing this book is to contribute to the reunification of evolutionary biology and Christianity on the latter related questions, and to the closing of the epistemological and ontological gaps that have presented a challenge to both theology and philosophy since the scientific acceptance of evolution. My main purpose, however, is to show that the achievement of synthesis between the two disciplines on these troublesome issues will provide corroborative empirical evidence for Christian teachings on the related issues of human origins, the origin of evil, and the existence of a beneficent deity. Synthesis between evolutionary biology and Christianity will also illuminate the relationship between scientific truth and religious truth overall; the result will be a deepened understanding of the meaning of truth generally and a new definition of progress that is inspired by both religion and science.

The key to resolving the apparent conflicts between religion and science—and especially between evolutionary biology and Christianity—lies, as I will show, in confronting the implications of evolution for morality; only when this has been achieved will it be possible to achieve synthesis between the two disciplines. This is because it is not the fact of evolution per se that poses a direct threat to religion, since creation through evolution in no way contradicts the theistic concept of an all-powerful, beneficent deity such as is portrayed in Christianity and also in the other two Abrahamic faiths of Judaism and Islam. It is the suggested processes of evolution that are problematic. The optimistic Lamarckian view, which in its contemporary secular form asserts a humanity that is evolving morally as well as technologically toward perfection, is a threat to religion generally since it removes any need for a transcendent dimension to human existence. The more pessimistic Darwinian view constitutes a challenge to all three of the Abrahamic faiths since it implies that the evolution of evil was inevitable. This casts serious doubt on the theistic view of God and also undermines the related belief that moral evil is a result of free human choice, the "original sin" of the first humans: it is the problem of evil in a new guise. The *problem of evil* refers to the question of how the suffering and evil that bedevil the human species can be reconciled with the existence of an omnipotent, omniscient, beneficent deity.

If God is all-powerful, all-knowing, and all-good, why did he create such a world? Christianity answers this question with the doctrine of a primal fall. God gave human beings free will, and they chose to disobey him, hence unleashing the forces of sin into the world. During the course of this book, I will argue that the implications of evolution for human behavior are of the utmost relevance to the problems that remain a barrier to full reconciliation between science and religion.

Chapter 1
Religion, Evolution, and Morality

The most exciting debate in morality today is taking place outside of its usual locations in philosophy and theology. It is from the field of science, more specifically from evolutionary biology, that radical new insights into human behavior are now emerging. The debate began at the start of the nineteenth century with the evolutionary hypothesis of Jean Baptiste Lamarck, a leading biologist of the era. It gained traction five decades later when the codiscoverers of natural selection, Charles Darwin and Alfred Russel Wallace, disagreed over the impact of evolution on human nature. The controversy quickly gained a philosophical dimension through the influence of evolutionary theory on leading thinkers such as Karl Marx and Friedrich Nietzsche, before losing its momentum during World War II and the postwar period. However, due to the achievement of synthesis between Mendelian genetics and Darwinian evolutionary theory that occurred in the 1930s, the disputed issues dramatically reemerged in mid-twentieth century science as memories of the war began to fade.

The contemporary controversy is generally associated with the opposing evolutionary perspectives of Richard Dawkins and

Stephen Jay Gould. Dawkins is the leading representative of Charles Darwin's contemporary disciples—neo-Darwinists—who argue that morality has evolved through natural selection and is genetically determined. Gould is a latter day advocate of progressive evolution through culture and is the most well-known modern disciple of Jean Baptiste Lamarck, whose followers are known as neo-Lamarckists. They assert, along with Lamarck, that morality evolves through culture rather than biology. It is a controversy that barely registers religiously, however, despite the ongoing conflict between science and religion on the subject of evolution. Institutional Christianity has tended to ignore the intricacies of the scientific debate, along with the scientific evidence and is generally doctrinally evasive on the issue. For evangelical Christianity, especially in the United States, the debate is especially controversial, since many sects interpret the Bible literally and hold to direct creation of the cosmos and humanity.

Interestingly, the contemporary controversies on the implications of evolution for human behavior both within science and between religion and science reflect the contrasting reactions of Darwin and Wallace, back in the nineteenth century, to their mutual discovery of natural selection. Darwin, who had originally intended to become an Anglican clergyman, famously lost his faith when confronted by the immense amount of suffering he believed to be caused in the world by selection-driven greed and aggression. If humanity had been shaped into a cruel and violent species by a process over which it had no control, was it still possible to believe in a beneficent deity? Wallace, on the other hand, began from a position of atheism and grew to believe that there had to have been benign divine intervention in the evolutionary process. The contrasting reactions of both men to their mutual discovery was, as we are about to see, due to differences in temperament as much as anything else.

Darwin's Pessimism: The Cruelty of Nature

Unlike some of his twenty-first century disciples, Darwin did not reject religious belief because natural selection removes the need to believe in a divine designer (in theology known as "the argument from design"). This was an argument that he had assimilated while preparing for holy orders at Cambridge, the university which, along with Oxford, had been central to the established Anglican faith since the Reformation. At Cambridge, science was described as "natural philosophy" to demarcate it from the more supernatural concerns of theology, and it was perceived as a subdiscipline of rational Anglican Christianity. The resulting alliance gave rise to a dynamic theology of nature known as "natural theology," a tradition developed to its highest level in William Paley's influential book of the same name. Paley's works were an important part of Cambridge's syllabus and were a major influence on Darwin as an undergraduate and as an evolutionary thinker because they stressed the evidence of God's designing powers in the adaptations found in nature. He became familiar with Paley's argument that

> The world was so full of design there must be a designer in the same way as a watch found on a path indicated the existence of a watchmaker (rather than a negligent owner). All of the "perfect adaptations" or "contrivances" [found] in nature were correspondingly understood as features specially created by God to fit each and every being for its role in an overall plan; a plan . . . characterized by stability, by inbuilt

> hierarchies, and by orderly arrangements reflecting the social and moral structure of Britain.[4]

This theological explanation for design in the world became the touchstone against which Darwin would test his future evolutionary hypotheses as he struggled to envisage an evolutionary mechanism that could explain the appearance of design. His eventual conviction that the sculptor of life on Earth is a mindless process such as natural selection did not, however, imply that there could not be a divine designer at one remove. Darwin was well aware of the theological counterargument against claims that creation through evolution dispenses with an important argument for the existence of God: this is that God, if he exists, is perfectly capable of working through natural processes—or secondary causes—such as natural selection. Indeed, many of his closest friends and supporters, including the great geologist Charles Lyell and the noted biologist Asa Gray, managed to maintain both their religious beliefs and their acceptance of natural selection by envisioning evolution as a process used by God to indirectly achieve his aims. Why then was Darwin so adamant in his rejection of a beneficent God and, ultimately, of any spiritual dimension to human existence?

I believe that the reasons for such radical disillusionment were almost inevitable, given his family circumstances, physical constitution, and fearless intellect. Years of observation had convinced him of the cruelty of nature. Natural selection seemed random and ruthless; those organisms that happened to be well-adapted to particular environments and circumstances

---

[4] Janet Browne, *Voyaging* (New Jersey: Princeton University Press, 1995), p. 129.

would survive at the expense of weaker, less well-adapted ones. Darwin was sensitive and squeamish; he had earlier given up medical studies due to an inability to stomach the sight of blood. His youthful passion for the natural world seems to have been gradually eroded during middle age: "What a book a Devil's chaplain might write," he wrote to Joseph Hooker, head of the Royal Botanic Gardens, "on the clumsy, wasteful, blundering, low and horridly cruel works of nature! . . . My God how I long . . . to wash my hands of it . . . for at least one long spell." In an 1860 letter to Asa Gray—at that time the professor of natural history at Harvard—he writes about his "painful" inability to see "as others do, & as I shd. wish to do, evidence of design and beneficence on all sides of us. There seems to me too much misery in the world. I cannot persuade myself that a beneficent & omnipotent God would have designedly created the *Ichneumonidae* with the express intention of their feeding within the living bodies of caterpillars, or that a cat should play with mice."[5]

Darwin's reference here to the predatoriness evident in the natural world hints at his growing conviction that human morality—including human predatoriness—has evolved through natural selection. During his *Beagle* years, he had encountered the practice of slavery and witnessed the brutality to which slaves were subjected. As a result he abhorred slavery and was duly appalled to discover a breed of red ants that captured and made slaves of other breeds. This would seem to suggest that in humans, as in ants, the propensity for slave-making might be innate, having evolved

---

[5] See Browne, *Darwin: The Power of Place* (New Jersey: Princeton University Press, 2002), p. 66; Frederick Burkhardt, Samantha Evans and Alison Pearn (eds.), *Evolution: Selected Letters of Charles Darwin 1860-1870* (Cambridge: Cambridge University Press, 2008), p. 11.

through natural selection. If, on the other hand, the human species was specially created by God, the situation was even worse, since God would be culpable of deliberately designing a horrific human practice. Darwin preferred to believe that such cruelty had nothing to do with an all-powerful deity. As he wrote to one of his many correspondents: "it has always appeared to me more satisfactory to look at the immense amount of pain & suffering in the world, as the inevitable result of the natural sequence of events, i.e. general laws, rather than from the direct intervention of God though I am aware that this is not logical with reference to an omniscient Deity."[6] Darwin's refusal to believe that God could be directly responsible for such cruelty seems ultimately to have extended to the notion that a beneficent God could permit it even indirectly through a process such as natural selection.

Darwin's observations on cruelty and suffering refer to "natural evil" and "moral evil," two categories to describe evil employed in philosophy of religion. The former relates to human suffering caused by natural disasters and illness, while the latter relates to suffering deliberately inflicted by human beings on one another: "man's inhumanity to man," so to speak. The existence of natural evils would not, from a theological point of view, rule out the existence of a good God; the existence of moral evil inevitably caused by evolution would. This is why Darwinism poses such a serious challenge to theistic religions such as Judaism, Christianity, and Islam; theism—as we saw in the Introduction—is the belief that God is all-good, all-powerful, and takes a personal interest in his creation and in each one of us. In his explanations to various correspondents as to why he cannot believe in a benign God, Darwin's references to natural evil outweigh those he makes

---

[6] Burkhardt, Evans and Pearn, *Evolution*, p. 156.

to moral evil. Quite apart from the sensitivity and squeamishness referred to above, Darwin's emphasis on natural evil was no doubt also influenced by the fact that he suffered for most of his life from chronic ill-health, which turned him into a virtual recluse. As a family man who idolized his children—three of whom died at an early age—he also agonized over their delicate constitutions; crises of health were a regular occurrence among the younger Darwins. Most devastating of all was the death at nine of his eldest daughter and favorite child Annie, a dreadful blow from which he never fully recovered. Darwin blamed himself for the delicacy of his children's health, fearing that they were at a double disadvantage in inheriting his own weak constitution through both their parents: Darwin and his wife Emma Wedgwood were first cousins. Viewing nature as he did through the dark prism of natural selection, Darwin must have felt that he and his children belonged to the ranks of the weak, at least where physical strength and stamina were concerned. It is my opinion, however, his emphasis on natural evil notwithstanding, that it was Darwin's observation of the impact of natural selection on human morality that was most decisive in his loss of faith. In response to a sensitive review of the *Origin* by his niece Julia Wedgwood, Darwin wrote:

> I think you understand my book perfectly, and that I find a very rare event with my critics . . . owing to several correspondents I have been lately led to think, or rather try to think over some of the chief points discussed by you. But the result has been with me a maze—something like thinking on the origin of evil, to which you allude. The mind refuses to look at this universe, being what it is, without having being

> designed; yet where one would most expect design, viz. in the structure of a sentient being, the more I think on the subject, the less I can see proof of design.[7]

It is interesting that in the first edition of *Origin*, Darwin placed the following quotation from William Whewell, a philosopher of science, on the flyleaf: "But with regard to the material world, we can at least go so far as this—we can perceive that events are brought about not by insulated interpositions of Divine power, exerted in each particular case, but by the establishment of general laws."[8] At first, Darwin was willing to concede and probably still believed that natural selection was one of the "general laws" referred to by Whewell, by means of which God creates. Over the years, however, as he refined his theory, Darwin became increasingly convinced that the need to compete for resources during times of crisis would have led, in the one species on Earth that had developed rational powers, to the evolution of calculated cruelty, aggression, and greed. He lost faith in the theistic God depicted in Christianity as he confronted the possibility that evil might be an inevitable result of evolution by natural selection in the human species. Although he publicly described himself as an agnostic, his biographer Janet Browne believes that ultimately Darwin slipped into atheism.[9] He would not be alone in his depressing conclusion; in the years following the publication of his theory two great philosophers who were at the peak of their powers in the nineteenth to early

---

[7] Browne, *The Power of Place*, p. 145.

[8] Francisco J. Ayala and Robert Arp (eds.), *Contemporary Debates in Philosophy of Biology* (Malden, MA: Wiley-Blackwell, 2010), p. 354.

[9] Browne, *The Power of Place*, p. 145.

twentieth centuries grasped the implicit nihilism of Darwinism in a way that even Darwin himself did not: these were Friedrich Nietzsche and Sigmund Freud, and they have been followed in the twenty-first century by others such as John Gray and Daniel Dennett. Indeed, if Darwin's speculations as to the mechanism that drives evolution are ever finally and fully proven to be true, they would imply either no deity at all, or necessitate a reversion to some sort of Manichaean dualism. Manichaeism as a religion explained evil as one of two main principles governing the world: evil is a positive force embedded in matter, while its opposite, good, belongs to the soul and the spirit.

In later chapters, I will give detailed consideration to Darwin's careful analysis of the relationship between animal and human morality; by the time he wrote *The Descent of Man*, which was published eleven years after *Origin*, Darwin had come to firmly believe that human morality had evolved solely through natural selection. As someone who prided himself on being a seeker after truth, he did not balk at the implications of what he had come to believe, although it removed all hope of reunion with his beloved Annie in an afterlife.

Wallace's Optimism: A Wonderful World

Wallace was in Indonesia during the aftermath of *Origin*'s publication but had been kept informed by Darwin of the controversies it provoked, in particular the encounter between Huxley and Wilberforce and the scorn being heaped by the developing and newly bullish scientific establishment on any scientists who espoused creationist ideas. This must have given Wallace pause for thought as, despite the fact that he would become Darwin's most courageous and staunch supporter, he had come to believe that there had to be a supernatural input into the development of the human species. Wallace's belief in a spiritual dimension to existence marked a surprising about-face from a longstanding atheism, and—in direct contrast to Darwin—had been caused by his discovery of natural selection. Though never conventionally religious as an adult, Wallace advanced scientific arguments that are far more rigorous than those of contemporary intelligent-design creationists for his claim of supernatural input into the evolutionary process: his belief in a supernatural dimension to evolution also made him far more optimistic than Darwin on the possibilities for human moral progress. So great were the hostilities that had opened up between the scientific and religious factions even at that early stage, however, that Wallace's well-meaning efforts to reconcile science and religion were ridiculed, while his own standing in professional science was seriously damaged. Indeed, were it not for the support of Darwin, Wallace's reputation might have been fatally wounded. Darwin's loyalty toward Wallace and his enduring friendship for the younger, less well-connected man is considered to be one of the most heartwarming in the history of science, all the more remarkable considering that Darwin's discovery of natural selection affected him in a way radically different from how it affected Wallace. Yet the disagreement between the two men over the implications of natural selection for the human condition was a major one. In his autobiography, Wallace tries to explain their differences thus:

> On this great problem the belief and teaching of Darwin was, that man's whole nature—physical, mental, intellectual, and moral—was developed from the lower animals by means of the same laws of variation

> and survival; and, as a consequence of this belief, that there was no difference in kind between man's nature and animal nature, but only one of degree. My view, on the other hand, was, and is, that there is a difference in kind, intellectually and morally, between man and the other animals; and that while his body was undoubtedly developed by the continuous modification of some ancestral form, some different agency, analogous to that which first produced organic *life*, and then originated *consciousness*, came into play in order to develop the higher intellectual and spiritual nature of man.[10]

There were various reasons for the divergence of view, one of which was undoubtedly a difference in temperament. Unlike Darwin, the gentleman naturalist aboard the *Beagle*, who had a fleet of helpers (crew and locals) to do the servile work on his land excursions and protect him from danger, the younger Wallace did not have a wealthy father to pay his way and had to rough it in wild, harsh, and unexplored terrain when he set out on his own travels on the Amazon and in the Malay archipelago. His many serious brushes with death—through tropical fevers, intrepid exploration, and shipwreck—toughened him up; he was in no way squeamish and never lost his joy in the natural world nor his zest for travel. Wallace was also highly sensitive to the fragility of nature, often described as an ecologist ahead of his time. This may have made him more aware than Darwin of the necessity for predation and death

[10] Alfred Russel Wallace, *My Life*, Vol. 2, p. 17. Cited by Andrew Berry (ed.), *Infinite Tropics* (London: Verso, 2002), p. 175.

in the maintenance of, as he put it, the "nicely balanced relations of organic and inorganic nature";[11] hence Wallace was not at all fazed by the occurrence of "natural evils" and did not, apparently, consider them to be of any relevance to the existence or otherwise of a benign spiritual domain. Indeed Wallace's enthusiastic admiration for the wonders of natural selection seems to have given him an exhilarating impression of the deity. The most important factor in his and Darwin's disagreement over natural selection and its implications for humanity was, however, Wallace's deep, first-hand knowledge of primitive tribes; it is doubtful if any other naturalist of the time had had so much exposure to "savage" peoples, as they were then unapologetically described. Wallace spent a lot of time living among and working with native tribes during his years of exploration, and—though he experienced the cruel and brutal side of tribal life—treasured his inspirational experiences with many of the native peoples he had encountered in the jungles of South America and Southeast Asia. In the course of his travels in the Malay Archipelago, he lived for a while among the Dyaks of Borneo, where "he was able to observe and admire [their] way of life, and to note the wonderful use they made of the plants and trees around them." Wallace also marveled at the Dyaks' honesty and the trusting atmosphere that resulted. During his earlier travels in South America he had been equally impressed by the work ethic and sense of community that characterized the Indians of the rainforest: in the village of Javíta, on the banks of the Orinoco in Venezuela, Wallace came upon what he considered to be a model of ideal community life. He observed the hard work and spirit

---

[11] Peter Raby, *Alfred Russel Wallace: A Life* (London: Pimlico, 2002), p. 122.

of cooperation among the inhabitants of Javíta who, once a week, would hoe and weed their village together. The community derived its income from cutting piassaba in the forest that was exported in its natural state or used in the manufacture of cables. Because of their favorable location, the Indians of Javíta were able to trade with the rest of the world without being exploited and were largely self-governing. Primitive humans are also spared, he felt, the type of competitive pressure that afflicts the "civilized," who must struggle all their lives to earn a living. Yet Wallace acknowledged "the joys, the pleasures and delights"[12] of civilization and the cultivated mind, pleasures unknown to more primitive races. The problem for Wallace was that such pleasures are bought at the cost of the suffering of the masses and enjoyed by a privileged minority; millions of Europeans, he felt, lived a life both physically and morally inferior to the Indians of the Amazon. Wallace had no doubt that the pursuit of money and profit was the cause of the misery; yet if the civilized races were to embrace the simple life in terms of material things (as Wallace himself would do

[12] Raby, *Wallace*, pp. 108–9, p. 123, pp. 71–72. The botanist Richard Spruce, an intermittent guest and traveling companion of Wallace during his years on the Amazon, experienced the Indians differently, barely escaping with his life on one occasion in the very area idealized by Wallace: "A country without priests, lawyers, doctors, police and soldiers, is not quite so happy as Rousseau dreamt it to be; and this, in which I now am, has been in a state of gradual decadence ever since the separation from Spain, at which period . . . the inhabitants rid themselves of these functionaries in the most unscrupulous manner." Writing about his experiences in *Travels on the Amazon* some years later, Wallace was aware that his own idyllic experience may have been a lucky and temporary one; yet for him the model of community life he had observed in the forest among the Amazon Indians convinced him that an earthly Utopia was possible.

throughout his long life) and cherish nature and cultural pursuits more, the problems of civilized lifestyles could be greatly lessened. In many respects, he felt, human nature in the wild is superior to the civilized version, and "progress" could mean looking to the past as well as the future.

Wallace was also an expert on the habits of the great apes and had actually reared by hand a baby orangutan whose mother he had shot. He was uniquely positioned, therefore, with his unmatched knowledge of, on the one hand, the "highest" representatives of the animal world and, on the other, the "lowest" (least civilized) representatives of the human, to speculate on the ape-human gap. He simply didn't believe that natural selection could have bridged the gulf between the ape and human species either intellectually or morally: "Natural Selection could only have endowed savage man with a brain a little superior to that of an ape, whereas he actually possesses one very little inferior to that of a philosopher."[13] Crucial to Wallace's conviction that there has to be more to human evolution than just natural selection were his reasonable, logical, and empirical observations on what natural selection can and cannot achieve. In a volume devoted to the limits of natural selection in its impact on the human species, Wallace argues that "savages" are mentally over-endowed for the primitive lives that they lead: while they have the same brain size and capacity as "civilized" humans, they have no use for the higher forms of intellectual activity that characterize more advanced societies. He points out that

---

[13] Wallace, *Contributions to the Theory of Natural Selection: A Series of Essays* (London & New York: Macmillan & Co., 1870), pp. 332–360. Cited in Berry, Infinite Tropics, p. 206.

> The power of conceiving eternity and infinity, and all those purely abstract notions of form, number and harmony, which play so large a part in the life of civilized races, are entirely outside of the world of thought of the savage, and have no influence on his individual existence or on that of his tribe. They could not, therefore, have been developed by any preservation of useful forms of thought. . . . But, on the other hand, we find out that every one of these characteristics is necessary for the full development of human nature. The rapid progress of civilization under favourable conditions, would not be possible, were not the organ of the mind *prepared in advance*, fully developed as regards size, structure, and proportions, and only needing a few generations of use and habit to co-ordinate its complex functions.[14]

Here Wallace makes the logical argument that it is impossible for natural selection to produce traits that are not functional within relevant environments; he is also hinting at what he believes to be the source of such elevated and yet intrinsically human faculties. For just as surely, he felt, as we can trace the action of natural laws in the evolution of organisms, so also can we simultaneously track the action of some unknown higher law responsible for what the natural processes could not have achieved. Wallace remarks that any inquiry into such a higher law is as "thoroughly scientific and legitimate as that into the origin of species itself. It is an attempt to solve the inverse problem, to deduce the existence of a new power of a definite character, in

[14] Ibid., p. 207. My italics.

order to account for facts which according to the theory of natural selection ought not to happen." For Wallace it was the very fact that the same great laws that had caused the evolution of all life were so obviously active in the evolution of humanity that provided evidence of a "Power" greater than natural selection that was also involved; this higher power or "Overruling Intelligence" has, he suggested, "watched over the action of those laws, so directing their variations and so determining their accumulation, as finally to produce an organization sufficiently perfect to admit of, and even to aid in, the indefinite advancement of our mental and moral nature."[15] Such purposeful activity was, for Wallace, analogous to the human activity of artificial breeding in order to produce higher forms and strains of animals, such as the Guernsey milk cow or the London draft horse.

Wallace's assessment of the human moral faculties in relation to natural selection is just as plausible and interesting as his observations on the mental faculties. He was in no way blinded to the deleterious effects of natural selection on human morality and never suggested, as some contemporary evolutionary biologists do, that natural selection has affected biological but not moral evolution. He argues, however, that just as natural selection can account for the ability to perform rudimentary mathematics but not for the capacity to imagine ideal conceptions of space and time, so, analogously, can it account for the practice of kindness and honesty, but not for the "peculiar sanctity" and even mystical sense that can attach to such abstract moral concepts as truth, justice, and beneficence in even the most primitive tribes.[16] Wallace's

---

[15] Ibid., p. 194. See also Raby, *Wallace*, p. 245.

[16] Wallace, *Contributions to the Theory of Natural Selection*, pp. 332–360. Cited in Berry, *Infinite Tropics*, p. 207.

dualistic speculations on our higher moral faculties have interesting implications for the relationship between the material and the spiritual in moral decision-making: if he is correct, abstract concepts of truth and beneficence belong to the spiritual faculties, or soul; it is in the interplay between these spiritual faculties and the biologically grounded traits developed over millions of years by natural selection that moral decisions are forged. There are further implications of Wallace's dualism with regard to the distinction of body and soul; the most prevalent contemporary philosophical and theological views of this relationship are either that matter and spirit are two opposing poles of the same substance (dual aspect monism), or the similar position that matter is evolving into ever higher spiritual forms. We will return to this discussion in a later chapter.

Wallace's belief in a benign "higher power" was not, however, a return on his part to traditional religion but an alternative to it. Indeed, he may well have been the very first human being to catch a glimpse of the supernatural domain as a direct result of modern, rational, scientific hypothesizing; a glimpse, moreover, that was free from the accumulated preconceptions, superstitions, prejudices, and corruptions of any established religion. This could explain his developing sense that conventional religion was something very different to what he thought of as the real thing. Wallace's youthful rejection of religion had been linked to his embrace of socialism which, it seemed to him, could be a real force for improvement of the human lot; established religion, he felt, was just another tool of repression of the masses. Nevertheless, Wallace did allow that orthodox religious belief as it had developed in civilized societies up to his own era had acted as a restraining influence on human behavior. Such religious belief was, however, being gradually eroded by the rise of science as a discipline in its own right; he began to fear the ghastly prospect of might constituting right on a huge scale in a world dominated by "the unbridled passions of the strongest and most selfish men." The only way to avoid such a scenario was to find forms of belief in the supernatural that could be proven scientifically and thus guarantee hope in an existence beyond this one:

> If all men, without exception, come to believe that there is no life beyond this life—if children are all taught that the only happiness they can ever reach is certainly limited to their lives upon this earth,—then, it seems to me, the condition of man would be altogether hopeless, because there would cease to be any adequate motive for justice, for truth, for unselfishness, and no sufficient reason could be given to the poor man, to the bad man, or to the selfish man, why he should not systematically seek his own personal welfare at the cost of others.[17]

Wallace felt that in the future, as the masses became more accustomed to the empirical findings and achievements of science, religious belief would only survive if religion too subjected itself to the rigor of scientific criteria for valid knowledge. He became convinced that a combination of scientifically enlightened religious beliefs and politics would usher in a new era of Lamarckian-style evolution through a combination of spirituality and culture.

It must be remarked at this point that Wallace's convictions concerning supernatural agency in the evolutionary process do not equate to the standpoint of contemporary intelligent-design

---

[17] Raby, *Wallace*, p. 245.

creationism. Its advocates share his motivation but not his methodology, since they blur the boundary lines between religion and science. Advocates of intelligent design envisage a "God of the gaps," who intervenes in evolution if the natural processes are incapable of producing the required organ or part of an organism. Divine agency, in other words, is intrinsic to biology and at times indistinguishable from it. Wallace conceives of a higher power that in some way acts as an overall guide to what is a random and even haphazard process, while remaining apart from it; any intervention that has occurred relates to the spiritual dimension—or soul—of the human being. Hence this power becomes partly intelligible in certain instances regarding the human species alone: it corresponds to the deity behind the "general laws" mentioned by Whewell and Darwin, and to the action of divine providence as proposed in the theology of, for example, Thomas Aquinas. For Aquinas, God, as prime mover, acts in the world through secondary causes or a great chain of events that is initiated and guided by him. Unlike the intelligent-design hypothesis, Wallace's hypotheses concerning the limits of natural selection with regard to the human species are, and will continue to be, subject to verifiability and falsifiability, as all truly scientific hypotheses must be.

The Philosophical Impact: Marx, Nietzsche, and Freud

The disagreement between Wallace and Darwin was soon reflected philosophically in the disparity that emerged between Karl Marx on the one hand and Sigmund Freud and Friedrich Nietzsche on the other on the subject of evolution. Marx had been influenced by Hegel, a leading theistic eighteenth-century exponent of German philosophy who, in typical eighteenth-century fashion, considered evolution to be a force that drives progress. Following Hegel, Marx held that evolution is progressive, rational, and directed toward an ultimate state of perfection. Unlike Hegel, however, he was not a theist and had to find something other than a divine explanation for his conviction that humanity would eventually be perfected by evolution. Like many atheistic philosophers seeking a secular narrative of progress, he was attracted to Darwin's theory of evolution by natural selection, becoming something of a philosophical hybrid in his approach to evolution. Marx transferred Darwin's competitive "struggle for existence" from the biological world to the sphere of human economics; he agreed with Darwin's theory that change over time comes about through conflict. However, as distinct from Darwin, who focused on conflict between individuals, Marx identified class struggle as the principle of historical change. Crucially, progress comes about through individuals *cooperating*—as opposed to competing with one another—albeit that such cooperation can be preceded by violent struggle against the oppressor class. Marx further argued that, after a revolution, the victorious class would work together peacefully in the pursuit of a more equal society; he was convinced that socialism would inevitably triumph as a result of the irresistible movement of historical forces. As Bertrand Russell has remarked, however, this philosophical optimism implies a belief in progress as a universal law and a level of optimism that "only theism could justify."[18]

Nietzsche and Freud, in contrast, were undiluted Darwinists who, as I have already pointed out, immediately grasped

[18] Bertrand Russell, *History of Western Philosophy* (London: Routledge, 1996), p. 754.

the nihilistic implications of Darwinism. Both men are, I believe, the philosophers par excellence of Darwinism, and each has written extensively on its implications for human behavior. In the aftermath of the publication of *The Descent of Man*, Nietzsche, for example, dramatically announced "the death of God." Because the emergence of human beings could now be explained by natural rather than supernatural forces, and since morality was not a faculty implanted in us by God but had, rather, evolved by way of our animal ancestors, there could no longer be any excuse for comforting ourselves with fairytales of a beneficent deity or of any deity at all. The most significant conclusion he reached, however—and the one most relevant to the overall argument of this book—is that there can never be progress in the moral sphere, since evolution has shown morality to be relative; hence human nature cannot change for the better, as Marx claimed. This is because of what he termed, paraphrasing Darwin, "the will to power." Nietzsche identified the will to power as the main motivator of human behavior: "Life itself is *essentially* appropriation, injury, overpowering of what is alien and weaker, suppression, hardness, imposition of one's own forms, incorporation and, at the least and mildest, exploitation."[19] The will to power ensures a constant struggle to achieve power and maintain it; whichever group holds power at any particular time will impose its norms in all significant areas of civilization—the arts, education, the media, the law, and morality—on those they rule. Inevitably, however, they will at some point be overthrown and the new rulers will impose *their* norms. Nietzsche felt that no form of rule could be considered

[19] Friedrich Nietzsche, *Beyond Good and Evil* (New York: Vintage Books, 1966), Sec. 259, p. 203.

better than another in any moral sense, though some might be preferable to others on a purely subjective basis. Nietzsche favored aristocratic rule on account of its encouragement of the arts and its broad-minded morality. Thus, while civilizations and societies change from culture to culture and from era to era, they don't progress morally; history, in fact, repeats itself in various ways.[20] Nietzsche termed his perspective on the dynamics that drive history "the eternal recurrence" and considered it to be the crown of his philosophical system.

Freud, also an atheist, grounded not only his philosophy but also his system of psychoanalysis on the view of human nature set out by Darwin in *The Descent of Man*, though it is the former which primarily concerns us here. Freud argued that the savage state persists under an outer veil of civilization and can reemerge in its raw and primeval form with extreme rapidity under certain circumstances, particularly during times of war. Against Marxism and Lamarckism generally, he also maintains that civilization is not a vehicle by means of which humanity can transcend its biological nature, arguing that even the highest cultural arts are linked to our most basic instincts, since they are made possible by the psychical displacement of energy (sublimation) that would otherwise be channeled into their gratification. Civilization, according to Freud, has two main tasks: it must protect men against nature and from one another, and for these reasons cooperation is necessary. The first task has been addressed with considerable success through technological

[20] Nietzsche's view of history emerges from his doctrine of "the eternal recurrence," which he introduced in his most famous work, *Thus Spake Zarathustra* (1883).

progress; the second—to progress morally—has been much less successful; this is because cooperation has to be enforced and is restricted to the bare minimum. More detrimentally, technological progress is all too often used as a tool for the satisfaction of the desires of the powerful, which entails the subordination of reason to instinct: the result is cyclical war and poverty. For Freud, human rationality is not powerful enough at the level of the species (he concedes that there can be individual exceptions) to overcome the strength of the instincts, and—like Nietzsche—he concludes that the world will continue to be marked by war, violence, and inequalities of wealth and poverty: "The ideal condition of things," he writes, "would . . . be a community of . . . men who had subordinated their instinctual life to the dictatorship of reason. . . . But in all probability that is a Utopian expectation."[21]

Overall, the nineteenth-to-early-twentieth-century philosophical reaction to Darwin decisively detached God from evolution, whether or not the philosopher concerned was optimistic or pessimistic regarding its impact on human nature. The philosophical denial of divine input into the evolutionary process was no doubt a factor in the reappearance of the original disagreement between Darwin and Wallace in mid-twentieth century science minus its theistic dimension.

Neo-Darwinism and Neo-Lamarckism

That early disagreement between the codiscoverers of natural selection vanished from scientific discourse until it reappeared,

---

[21] Sigmund Freud, *Civilization, Society and Religion* (London: Penguin, 1991), pp. 353–360. See also pp. 251–339.

in primarily moral form, in the 1960s. The contemporary version of this debate is largely due to advances in scientific understanding of how behavioral characteristics are inherited. In his lifetime, Darwin had been able to explain how traits are "selected" but not how they are passed on down through the generations. Unknown to him, Gregor Mendel, an Austrian monk, had discovered the basic principles of heredity in the 1860s. Mendel experimented with crossing varieties of peas in relation to color and shape of seed, counting all types and combinations in the offspring for several generations. His experiments showed that characteristics are transmitted to offspring via the genes of their parents and allowed him to deduce the statistical laws that would shape the foundation of modern genetics. Another significant development occurred in 1882 when the German cytologist and zoologist August Weismann demonstrated the impossibility of Lamarck's claim that beneficial characteristics, developed during a person's lifetime, can be passed on to his or her children: as we saw in the Introduction, one of Lamarck's main tenets was that acquired characteristics can contribute to progress through inheritance. Weismann showed that there is a barrier between the "soma," or body cells of an organism, and the "germ plasm," or sex cells. The implications for heredity are that even if an organism acquires new abilities in response to environmental conditions, such changes cannot be transferred to the sex cells (ova and sperm); only the material contained in the sex cells is involved in heredity. An important implication of the latter discovery for behavior is that moral characteristics that humans—at either the individual or group level—may develop during their lifetimes cannot be inherited biologically by their descendants. Although the up-and-coming discipline of evolutionary developmental biology has shown that in certain cases environmental factors can bypass the genes themselves by influencing gene expression and that the resulting changes—particularly if related to health—can sometimes be passed on to offspring,[22] this is not considered to be a significant challenge to the neo-Darwinian view of behavior. It is argued that natural selection itself has molded developmental processes so that they can behave in different ways in response to different environments, hence increasing the plasticity of organisms and their ability to adapt to a variety of environments.

Darwin's and Wallace's theory finally developed full status as a scientific discipline in the 1930s, when synthesis between Darwinian evolution and Mendelian genes was begun by R. A. Fisher in *The Genetical Theory of Natural Selection*. Fisher, in

---

[22] It has been shown, for example, that daphnia water fleas, which develop protective spines in the presence of predators, pass them on to their offspring. It has also been shown that in the offspring of frail, yellow cancer-prone mice called agouti mice, so-called because they carry a gene (called the agouti gene), which predisposes them to cancer, the gene can be switched off in the offspring if the mothers are fed a diet that is rich in chemicals called methyl-donors. The offspring are then born brown and healthy. In the case of humans, an experiment that studied crop data and population over the course of two centuries in the isolated Swedish village of Overkallix found that the quality of teenagers' diets impacted the health of their grandchildren. Teenage boys brought up in times of famine, for example, had grandsons that lived longer, while teenage boys who lived in times of plenty tended to have grandsons who became diabetic. The effects were also gender specific, so that girls brought up in times of abundance had grandchildren who were less inclined to develop heart disease. The crucial point is, however, that epigenetic (literally, "outside the genes") inheritance leaves the genes themselves intact, merely altering their expression. Epigenetic inheritance is also, apparently, easily reversible and does not seem to persist for more than a few generations. See Fern Elsdon-Baker, *The Selfish Genius* (London: Icon Books, 2009), pp. 119–123. See also Michael Ruse and Joseph Travis (eds), *Evolution: The First Four Billion Years* (Cambridge: The Belknap Press of Harvard University Press, 2009), p. 898.

conjunction with J.B.S. Haldane and Sewell Wright, realized that the Mendelian laws of inheritance could not by themselves bring about a change in the average genetic composition of a population and that evolutionary change occurred through the phenomenon of adaptation sorting through and selecting Mendelian genes. Taken together, the work of these three pioneers gave rise to theoretical population genetics and the modern synthetic theory of evolution known as neo-Darwinism. The discovery in 1953 by James Watson and Francis Crick of the structure of the DNA molecule further clarified the role of genes in heredity and, taken together with the previous advances, precipitated a new wave of research on that part of Darwin's and Wallace's theory that had caused Darwin so much angst: the implications of natural selection for human morality. The emphasis of postwar neo-Darwinism was not, however, on the evolution of cruelty but on the evolution of cooperation, a phenomenon that had puzzled Darwin himself. In the 1970s a new evolutionary discipline known as sociobiology emerged, founded by E. O. Wilson. His brilliantly original, if extremely controversial, *Sociobiology: The New Synthesis* fearlessly applied genetic theory first to the study of animal social behavior and then to human social behavior. The assertion that cooperative behavior is genetically determined resulted from a breakthrough in understanding of how genes work that enabled second-wave evolutionary biologists to solve a conundrum posed by Darwin in *The Descent of Man*, when he asked:

> How within the limits of the same tribe did a large number of members first become endowed with [the] social and moral qualities, and how was the standard of excellence raised? It is extremely

> doubtful whether the offspring of the more sympathetic and benevolent parents, or those which were the most faithful to their comrades, would be reared in greater number than the children of selfish and treacherous parents of the same tribe. He who was ready to sacrifice his life, as many a savage has been, rather than betray his comrades, would often leave no offspring to inherit his noble nature. The bravest men . . . would on average perish in larger numbers than other men. Therefore it seems scarcely possible . . . that the standard of their excellence . . . could be increased through natural selection, that is, by the survival of the fittest.[23]

Darwin, whose theory was based on the hypothesis that natural selection acts on the most competitive and able individuals in any group, was puzzled as to how qualities such as self-sacrifice in battle could have evolved. The fact is that, as many neo-Darwinists point out, cooperation abounds in nature, and this would seem to contradict the theory of natural selection which, on a superficial level, seems to imply that the most aggressive and selfish individual organisms must be the most successful. Yet African kingfishers will feed and protect young that are not their own, and worker honey bees will sacrifice themselves to protect the hive when they sting an intruder. Worker bees are also sterile, having evolved to sacrifice their own reproductive potential to devote themselves to the welfare of the hive. In accordance with a strict theory of natural selection, a stinger on

[23] Darwin, *The Descent of Man*, Vol.1 (Princeton: Princeton University Press, 1981), p. 163.

a bee should not have evolved, nor should sterility.[24] As a result of these anomalies, evolutionary biology had been driven in the postwar period by the idea that, in social species such as our own, natural selection favors traits that increase group survival. In other words, in the case of morality, selection operates on the survival of the group rather than the individual, meaning that those groups that contained the greatest number of cooperative individuals would prevail over more selfish ones during times of conflict, thus increasing the general standard of morality. Group selection, however, can't explain how such cooperative behavior could have evolved in the first place. The first altruistic fully modern human individual would have been at such a great disadvantage in relation to the other individuals of his tribe that survival for long enough to reproduce his altruistic genes would have been highly problematic if not impossible. Another problem is that, during evolutionary history, there would have had to have been near constant warfare between tribes (on a par with individual competition and conflict within tribes) in order for natural selection to be able to "select" for cooperation at group level. The notion of group selection had, nonetheless, become widespread and popular both in science and among the general public; it portrayed moral evolution as driven by the unselfishness of individuals who act altruistically in the best, long-term interests of the group (or society) rather than in their own short-term interests, thus driving the species progressively forward. This Lamarckian-tinged version of evolutionary biol-

---

[24] See, for example, Ruse and Travis, *Evolution*, pp. 407-410; and Carmen Sapienza "Does Selection Operate Primarily on Genes?" pp. 123–140 in Francisco J. Ayala and Robert Arp (eds.), *Contemporary Debates in Philosophy of Biology* (Oxford: Wiley-Blackwell), 2010.

ogy was acceptable in a way that the survival-of-the-fittest version was not: after the horrors of the Second World War, which included Hitler's tragic misappropriation of Darwinism for eugenic purposes, there seems to have been a profound need to believe that the excesses of violence and barbarity it engendered were a one-off aberration. Indeed, the high-point of revulsion at the wars of the twentieth century was reached during the 1960s, when a whole generation of Western teenagers and young adults rebelled against the mores of their parents and grandparents. It was also during the 1960s, however, that a new generation of evolutionary theorists found a way to solve Darwin's conundrum as to how cooperation can flourish in a selfish world with the discovery of the evolutionary mechanisms of kin selection, reciprocal altruism, and evolutionarily stable strategies. It was the latter discoveries that would cause bitter and continuing division among evolutionary biologists on the extent to which human nature has been shaped by natural selection. At issue, of course, is the vital question of the extent (if any) to which human nature is genetically determined and the implications of determinism for the possibility of human progress. The result is that evolutionary biology has now split into two groupings: neo-Darwinism, which—as we will shortly see—attributes the previous role of God in evolution to a combination of natural selection and genetic engineering; and neo-Lamarckism, which substitutes cultural evolution for the deity.

The evolutionary mechanisms mentioned above—kin selection, reciprocal altruism, evolutionarily stable strategies and cultural evolution—are, therefore, of crucial importance not just to the internal scientific dispute but also to the related dispute between religion and science, and hence my main purpose in writing this book, which is to demonstrate that, contrary to popular opinion, they provide evidence for the existence of a beneficent deity. For this reason I must now subject them to detailed analysis. If you have read the many books published in recent years on popular science, it will be a case of revisiting concepts with which you are already familiar; the context within which they are being examined is, however, new, and the conclusions that I draw throughout the following chapters are original. If you are a person of strong religious faith who has avoided works of popular science up to now or who considers them to be irrelevant, I would urge you to put your reservations to one side and read what follows. I have written it particularly for those who fear that the genetic determinism advocated by evolutionary commentators from various disciplines is a threat both to the well-being of the human species and to faith itself. By the end of the book, I can promise that, at the very least, your preconceptions will have been challenged, and you may even be pleasantly surprised.

Chapter 2
Religion, Evolution, and Progress: The Concept of the Evolutionarily Stable Strategy

Although he might not agree, I have no doubt that another significant step in the reunification of religion and science was taken with the publication in the 1960s of a paper on kin selection by world-renowned theoretical biologist William Hamilton.[25] Hamilton's deceptively simple solution to Darwin's evolutionary dilemma was that self-sacrificial behavior can evolve if such behavior benefits relatives of the altruist, since they share many of the same genes: hence the term *kin selection.* The notion of kin-selection also gave rise to the concept of "inclusive fitness," which refers not just to individual fitness but also to the extent that individual fitness benefits offspring, siblings, and cousins. This means that when an organism sacrifices itself to save a relative or forgoes its own reproduction to help a relative reproduce, it is perpetuating its own genes in an alternative and more effective way through the genes that they share in common. As

[25] William Hamilton, "The Genetical Evolution of Social Behavior" (I and II), *Journal of Theoretical Biology*, Vol. 7 (1964), pp. 1–16, 17–52.

J. B. S. Haldane, another of the pioneers of contemporary neo-Darwinism has put it: "Would you lay down your life for your brother? For two brothers certainly, or four cousins."[26] During our evolutionary history, most people lived in small tribes and would have been at least distantly related to most other members of the tribe, hence facilitating the evolution of self-sacrificial behaviors. The implication is, however, that evolution occurs at the level of the gene rather than the individual, as Darwin had thought, and certainly not at group level. This would mean that behavior is even more determined than Darwin had envisaged, since the self-sacrificing individual is not making a purely rational choice but is motivated instinctively by a gene acting "selfishly" (as described by Richard Dawkins in *The Selfish Gene*[27]) to ensure that that individual's genes will continue to flourish in the gene pool. This does not, of course, detract in any way from the heroism of the person who is prepared to lay down his or her life for others or indeed from any other form of self-sacrificial behavior; what it does do is explain how altruism of a very admirable kind can be produced by natural selection.

In collaboration with theoretical biologists George Price and John Maynard Smith, Hamilton also worked on other groundbreaking concepts such as reciprocal altruism and evolutionarily stable strategies. Reciprocal altruism describes the 50 percent of cooperation found in nature that is not kin related; this form of

---

[26] Andrew Brown, *The Darwin Wars* (London: Simon and Schuster, 1999), p. 8. In such cases, the cost to the altruist is outweighed by the benefit to the recipient(s), who in total will perpetuate more of the altruistic individual's genes down through the generations than the individual himself or herself possesses. The formal mathematical calculation of this is known as "Hamilton's Rule." See Dawkins, *The Selfish Gene*, pp. 96–97.

[27] *The Selfish Gene* (Oxford: Oxford University Press, 1989).

quid-pro-quo morality evolved as the result of enlightened self-interest and eases our way in the world among those to whom we are not related. The argument goes that in species that are capable of recognizing and remembering each other as individuals, "associations of mutual benefit will evolve if each partner can get more out than he puts in." In as much as one behaves in a friendly and altruistic manner toward others, he or she will be prepared to behave similarly toward you, and humans who are helped by others are more likely to survive and reproduce than those who are not. As a result of these theoretical breakthroughs, many evolutionary biologists now claim that altruism, as it is understood by religions such as Christianity, is an illusion, and that the whole notion of "morality" must be reassessed in light of developments in evolutionary theory:

> Our sense of morality is an adaptation. We survive and reproduce more effectively with it than we do without it. In the past, those people who lacked a sense of morality simply tended to be ostracized and at a disadvantage. So they failed to survive and reproduce as efficiently as those with a sense of morality.[28]

While the above statement may be true, there is no need to conclude that religious concepts of altruism must be an illusion. I would argue that the mechanisms of kin selection and reciprocal altruism simply illuminate aspects of how morality evolved; they are in no way a threat to theistic understandings of morality.

The same cannot be said, unfortunately, for the concept of the evolutionarily stable strategy (ESS), which is related to

[28] Michael Ruse, *The Darwinian Paradigm* (London: Routledge, 1989), p. 19.

reciprocal altruism: it is, in my opinion, the most important for morality of the three mentioned above. The ESS concept is based on the application of game theory to animal and human behavior. The moral significance of the ESS lies in its clarification of how genes are so proportioned and balanced in populations that all morally relevant behaviors are held in a sort of dynamic equilibrium with one another. Applied to human populations, this means that altruism can never increase sufficiently to defeat the darker human propensities such as aggression, greed, and cruelty. The concept of ESS helps us to understand how qualities such as peacefulness, cooperation, and even timidity can flourish in a ruthlessly competitive, selfish world. In *The Selfish Gene*, Richard Dawkins uses behaviors drawn from the natural world to illustrate how the various characteristics interact, and I will follow suit here, albeit using more biblically resonant examples.

Let's imagine that there are only two sorts of fighting strategy in the population of a given species, named *lion* and *lamb*. Lions are always predators (aggressors), while lambs are always peacemakers (altruists). In a population comprised totally of lambs, peace will reign until, inevitably, a mutant lion arises among the lambs. Lions always beat lambs in fights, and consequently lion genes spread throughout the population to such an extent that soon, the entire population consists of lions. Now there is permanent conflict, since lions always fight with one another and are evenly matched. At some point, however, a mutant lamb will be produced within the population. The lamb cannot win a fight and so always concedes, thus ensuring that it is never injured. The lions, meanwhile, suffer many injuries since they are bound to lose half their fights. Eventually, lamb genes begin to prosper in the population until finally the lions are vanquished and the population reverts to a population of lambs. Theoretically, there could be a continuous oscillation between lions and lambs; natural selection will however establish a stable ratio (an ESS) between the two fighting strategies, thus curtailing excessive aggression and maximizing environmental benefits. This ratio turns out to be five-twelfth lambs to seven-twelfth lions. Though the proportion of aggressors is slightly greater than that of peacemakers, the payoff in terms of environmental resources is exactly equal for both strategies. Therefore, selection does not favor either strategy above the other. The strategy is stable because if the number of lions in the population were to increase, lambs would gain the advantage and the ratio would swing back to the stable rate. There is no question, however, of the lions suddenly lying down with the lambs and becoming timid herbivores. Moreover, in applying "lion" and "lamb" strategies to the human situation, it is important to remember that, in reality, a variety of behaviors is involved, and that, in the case of humans, most individuals are a combination of lion and lamb.

The operation of reciprocal altruism is greatly illuminated if we invent two new strategies called *chimpanzee* and *bonobo* and substitute them for lions and lambs. We will then add a third strategy, called *elephant* to the mix. Let's suppose that, due to an environmental disaster, there is a serious shortage of food in our sample species. All of the animals must share their food with one another in order to survive. Bonobos always share their food, and chimpanzees always accept it while never sharing their own; as a result, there is a population explosion among the chimpanzees while the bonobos soon become extinct. As soon as the last bonobo has died, however, the chimpanzee population will also begin to drift toward extinction. It is at this point that the elephant strategy emerges; elephants share with strangers as well as individuals who have previously shared with them. If, however, an individual with whom they have shared reneges and refuses to share with them, elephants remember the incident and bear a grudge; they will refuse to feed chimpanzees in future. In a population of chimpanzees, bonobos, and elephants, the elephant strategy will ultimately win out. It is an ESS because in a population consisting largely of elephants, neither chimpanzee nor bonobo will be able to invade. As with the lion and lamb strategies, it is immune to treachery from within.

What these strategies illustrate is that in any given population each individual is trying to maximize his own success; therefore, in relation to the group, strategies must evolve that are immune to betterment by the treachery of any deviant individual. Game theory enables us to see how a collection of independent, selfish entities can appear to be a single, organized whole. According to Dawkins, however, populations actually have no group will or purpose that transcends individual self-interest, just a "pecking order" that allows the "fittest" individuals to achieve maximum power and wealth, while enabling the majority to coexist as comfortably as possible.[29] Individuals will act for the good of the group only if such actions will benefit them as individuals more than if they don't so act. Further, game theory shows that indiscriminately altruistic behavioral strategies are extremely disadvantageous in evolutionary terms; an individual who is altruistic in a nonstrategic way risks annihilation. Yet we can all see that if a majority of individuals was able

[29] Dawkins's application of game theory to the behavior of social species can be found in *The Selfish Gene*, pp. 67–87 and pp. 184–186. Game theory is the branch of mathematics used to illustrate the working out of evolutionarily stable strategies in populations. See also Brown, *The Darwin Wars*, and Hamilton, *Narrow Roads of Gene Land: Evolution of Social Behavior* (Oxford: Oxford University Press, 2006), pp. 373–385.

to put their short-term interests to one side when the welfare of society demanded it, everyone would be better off overall.

There is no doubt, however, that the gene's eye view of evolution is pessimistic where human morality is concerned. Its implications are that evolutionary forces impose such severe penalties on nonstrategic altruism, or individual self-sacrifice outside of strictly proscribed situations, that the possibility of the good in human nature increasing sufficiently to defeat the bad is problematic; in other words, war, conflict, and excessive inequalities in material goods may be permanent, ineradicable features of human existence. While human progress cannot be defined in moral terms alone of course, it is certainly more dependent on moral progress than on, say, technological progress or advances in medicine. It could even be argued that possibilities for human technological progress can only be fully realized within a context of cooperation, compassion, and social justice.

The disturbing implications for morality of neo-Darwinism have caused evolutionary biology to split into two opposed groups, one of which is associated with Dawkins—an ethologist as well as a science writer—the other with Stephen Jay Gould, a paleontologist who has written prolifically against the notion that natural selection has had a significant impact on human behavior. Gould, who started his career as a conventional neo-Darwinist, has challenged the neo-Darwinian paradigm by arguing that natural selection is not even the main mechanism of evolution. There is a huge literature on the science of the disagreement between both sides, which it is beyond the scope of this book to explore; in discussing the issues in terms of the well-known rivalry between Dawkins and Gould, I hope to streamline what is a complex and multifaceted argument while at the same time highlighting its most important points. It is also my opinion that the debate between the two men is a classic that has set and will continue to set the parameters of current and future debates on the same topic.

The Culture Wars

The role that culture plays in evolution is the question that most notably divides both factions in the Darwin disputes. Indeed, neo-Lamarckists are fond of pointing out that in dealing with the evolution of human morality, Darwin's strict individual selectionism wavered slightly on the issue of human morality and its relationship with culture. He certainly speculated that the social instincts that bind social species such as ours had brought about some evolution in human behavior. He frequently cites the values of civilized society, particularly his own Victorian society, as evidence of moral progress since prehistoric times. He also hoped that the social instincts would triumph as civilization advanced:

> As man advances in civilization, and small tribes are united with larger communities, the simplest reason would tell each individual that he ought to extend his social instincts and sympathies to all the members of the same nation, though personally unknown to him. This point once being reached, there is only an artificial barrier to prevent his sympathies extending to the men of all nations and races.[30]

Hopes and feelings aside however, Darwin's concession to group selection is heavily qualified. Most significantly, he offers an alternative to it by suggesting that since many members of small tribes are related, they would be more likely to help one

---

[30] Darwin, *Descent*, pp. 100–101.

another, and also that those who help others are likely to be helped in return: "As the reasoning powers and foresight of members became improved, each man would soon learn from experience that if he aided his fellow-men, he would commonly receive aid in return."[31] These suggestions anticipate the neo-Darwinist theories of kin selection and reciprocal altruism.

Darwin's doubts about the efficacy of individual selection in relation to human morality were undoubtedly due to his ignorance of the mechanism of heredity. Had he known about the way characteristics are transmitted down through the generations, he would have been able to understand how genes for altruism can spread through the population of social species. Dawkins believes that "selfish gene theory is Darwin's theory expressed in a way that Darwin did not choose but whose aptness, I should like to think, he would instantly have recognized and delighted in."[32] Given Darwin's overall individual selectionism, I am sure that he would indeed have agreed with the findings of Hamilton, Trivers, and Maynard-Smith among others, though I am not sure that he would have been "delighted" by them.

As a means of countering the pessimistic implications for human morality of selfish gene theory, Dawkins has introduced the concept of "memes." The word *meme* derives from the Greek word *mimema*, which means something that is imitated. He defines memes as being individual cultural units that propagate social memories and new ideas—including moral ones—in a way that parallels the spread of genes.[33] Cultural concepts and

[31] Ibid., p. 147. Darwin calls this "a low motive."

[32] Dawkins, *The Selfish Gene*, p. viii.

[33] Examples of memes, according to Dawkins, "are tunes, ideas, catch-phrases, clothes fashions, ways of making pots or of building arches." See *The Selfish Gene*, p. 249.

ideas compete with one another, and the most successful ones will travel from mind to mind as opposed to from body to body. In relation to moral progress, Dawkins has argued that knowledge is power; now that we know about our selfish genes and their deleterious effects, we are in a position to defy them through appropriately chosen memes as well as other methods that will shortly be discussed. For now, however, we turn to the neo-Lamarckian perspective on what constitutes cultural progress, whose most well-known representative is Gould.

Punctuated Equilibrium and the Role of Culture

The role that culture plays in evolution is the question that most notably divides both factions in the Darwin disputes. Due to Darwin's qualified concession to group selection where morality is concerned, Gould and many others in the field refuse to concede that Darwin's intellectual legacy belongs solely to Dawkins and his fellow selfish gene theorists, directing their attacks on neo-Darwinism and the reductive approach to the human person that it implies. Quite apart from the question of behavior and morality is the question of the human aesthetic sense: Is it possible or even feasible that all the artistic achievements of culture and civilization can be explained in terms of individual competitiveness? They argue that through cultural achievements in politics, education, technology, the arts, and morality, the human species—in distinction from all others—has evolved far beyond its biological origins.

In his book *Life's Grandeur*, Gould points out that adaptive or genetic evolution is Mendelian and therefore incremental and slow, while cultural evolution is Lamarckian and speedy:

> Any cultural knowledge acquired in one generation can be directly passed to the next by what we call,

> in a most noble word, education. If I invent the first wheel, my brainchild is not condemned to oblivion by hereditary impassibility (as any purely bodily improvements would be). I just teach my children, my apprentices, my social group, how to make more wheels. The point is so simple, yet so profound. Reading, writing, filming, teaching, practicing, apprenticing, learning—all the distinctly human activities that pass knowledge between generations—act as the Lamarckian boosters of our cultural history.[34]

Here Gould removes the emphasis on adaptive fitness as the most distinctive feature of the human species in order to distinguish between the amorality of evolution on the one hand and human purposefulness on the other. With regard to morality, he goes on to suggest that our moral faculties have evolved as a side-effect of being a social species. Gould terms such side-effects *exaptations*—phenomena that can then be coopted by natural selection for other purposes. The moral sense, he believes, is an exaptation that has been coopted by humanity as a species; moral progress occurs when we direct and guide our moral potential in the Lamarckian manner he describes. Moral progress, therefore, comes about through conscious human cooperative cultural endeavor as opposed to individual competitiveness, whether genetic or memetic. In the context of developing genetic technologies, Gould is notably wary of the whole

[34] Stephen Jay Gould, *Life's Grandeur: The Spread of Excellence from Plato to Darwin* (London: Vintage, 1992), p. 222. See also "More Things in Heaven and Earth" in Hilary Rose and Steven Rose (eds.), *Alas Poor Darwin* (London: Jonathan Cape, 2000), pp. 85–105.

notion of evolutionary progress, since it could open the door to eugenics. He prefers to talk about cultural change rather than cultural evolution. He also argues that an evolutionary mechanism he terms *punctuated equilibrium*—and not natural selection—is, in fact, the main mechanism of evolution. Punctuated equilibrium is associated mainly with paleontologists since it is based on the macroevolution visible in the fossil record as distinct from the microevolutionary focus of the gene's-eye view. Its supporters claim that the fossil record does not show gradual change but exhibits uniformity and stability over millions of years, with new species suddenly appearing. This would seem incompatible with the slow and continuous adaptation implied by natural selection; Darwinian processes would suggest that all species should be constantly evolving into their successors, and Darwin himself was adamant that natural selection is incompatible with sudden evolutionary leaps of any kind. While the fossil record does show many intermediate forms however, change frequently seems to occur across gaps or breaks with new species emerging in what Gould describes as a "geological moment." This apparently sudden emergence of new species had been previously attributed by evolutionists to an imperfect fossil record, which could not reveal the whole evolutionary story. Gould would also have been aware of the general consensus that gradual, imperceptible adaptive change might become suddenly visible under the precipitative force of geographical isolation in a new environment, especially when, as he points out, a geographical moment may represent many thousands of years of development. He has stood by his claim, however, that punctuated equilibrium is a primary pattern of evolution and that species' changes occur suddenly in one or two generations; he highlights the essential abruptness of evolution as opposed to its continuity, portraying a state of stasis or "equilibrium" that is suddenly "punctuated" by the emergence of new species. Over the years, Gould developed a multilevel, hierarchical view of evolution, in which the significance of adaptation is deemphasized. Natural selection continues to operate at the level of the individual, but it is just one evolutionary force among others. The assertion of evolutionary mechanisms other than natural selection is not in itself a challenge to the neo-Darwinian paradigm since other causes of evolution such as genetic drift and gene flow are well known.[35] The radical element of Gould's recalibration of evolutionary mechanisms is his claim that the most important patterns and dynamics of evolution operate at the

---

[35] Genetic drift occurs because, in accordance with Mendelian genetics, each individual receives two copies (alleles) of each gene from his or her parents and passes one on at random to their own offspring. For example, in any one family of four children, two should be boys and two girls; it not infrequently happens, however, that such a family might consist of three boys and one girl, three girls and one boy, four boys, or four girls. Theodore Dobzhansky gives another example of random genetic sampling when he describes an urn containing equal numbers of differently colored balls. If a blindfolded person draws a sample of balls from the urn and places them in a second urn, it is likely that the sample group of balls will differ in frequency in relation to color than those in the original urn. In terms of animal populations (including human ones), if a catastrophe of some kind occurs and leaves just a small group of survivors, the chances are that the composition of their genes will differ to those in the original larger population; this is known as genetic drift. A similar form of drift can occur when a small section of a population becomes separated from the main group by migration. In each case, the new genetic composition may result in the founding of a new species, even though natural selection has played a relatively incidental part in the process. Nonetheless, this form of evolution is still explained by Mendelian genetics. Gene flow is the opposite of genetic drift in that it describes large populations in which there are no significant geographical barriers and genes can flow unhindered throughout the population, thus preserving homogeneity and preventing speciation.

level of species; understood in the human behavioral context, this would contradict the argument of the opposing side that societies have no group will or purpose independent of individual self-interest. Moral action carried out on behalf of the species would amount to the most altruistic kind of behavior possible since anyone capable of it would be prepared to put the good of humanity above individual, tribal, or national interest at great cost to himself or herself and with no genetic payoff. It would imply a total severance of altruism from its biological roots. Most significantly, it would mean that at any particular time and in any particular population, there must be sufficient numbers of such altruistic individuals to reach a critical mass capable of transforming the morality of the society as a whole or of the species. In exceptional circumstances such as wartime and the aftermath of war, or heroic feats of exploration such as, for example, space travel, we do see altruism operating at the level of the group and the species. Indeed, space exploration has made it possible for behavior on behalf of the species to be consciously undertaken in a way never before possible. However, such self-sacrificial behavior is most likely limited to particular situations and circumstances, and is sporadic. In *Descent*, Darwin suggests that heroic self-sacrificial behaviors on behalf of the group probably evolved in response to the praise and glory accorded to those who performed such feats. The often remarked on "blitz spirit" of courage and cooperation during World War II and the sacrifices made by the German people as they worked to build up their defeated country would seem to indicate that in times of crisis, the long-term welfare of the group attains a clarity and focus that enables the masses to put the common good ahead of their own short-term individual interests. Catastrophes and disasters seem to bring out the best in human nature and permit empathy to flow unhindered. It is during times of peace that human greed and aggression create environments in which petty envies, jealousies, treachery, and deceit can simmer, restricting human altruism and even facilitating strategic altruistic display rather than the genuine article. I would go so far as to suggest that the immense ability for heroism and self-sacrifice that becomes visible at times of catastrophe or during feats of exploration demonstrates the greatness of our species and is a strong indication of a group will that is actually indicative of an underlying spiritual nature. However, the limitation of altruistic behaviors implied by the operation of ESSs would imply either that there cannot be enough altruistic behavior or altruistic individuals in a given society at any one time to reach the critical mass necessary for systematic, targeted moral evolution, or that genuine altruism is so penalized in most human situations that it cannot break through the barriers that power puts in its way.

In assembling his arguments for a multilevel, multi-faceted concept of evolution, Gould's main purpose seems to be to challenge the notion that human nature has been shaped in any significant way by natural selection and to provide an integrated view of cultural progress that includes moral progress; the claim that evolutionary jumps (or *saltations*, as they are termed by scientists) are characteristic of the fossil record allows him to claim that there are parallel leaps of cultural progress in the human sphere and hence to support his argument that the human species can progress rapidly through the cultural inheritance of acquired skills, characteristics, and moral behaviors. The Lamarckian implications for human morality of this scenario are, in fact, boundless. The right choice of a purposefully directed moral education could, for example, result in rapid moral evolution at the level of species. Gould's painstaking, meticulous, and brain-wracking efforts—especially in his final work *The Structure of Evolutionary Theory*[36]—to establish punctuated equilibrium as the most important mechanism of evolution are undoubtedly in part an admirable attempt to counter the claim that, due to genetic determinism, human moral progress is problematic.

If Dawkins, in focusing on the impact of natural selection on behavior is the anointed disciple of his idol, Darwin, Gould—though he might be loath to admit it—is reminiscent of a latter-day Wallace, minus the theism. Wallace also believed that the higher human faculties—aesthetic and moral—cannot be attributed to natural selection alone. Wallace, of course, was able to maintain a rigorous doctrine of natural selection, which included its impact on morality, due to his belief in a spiritual dimension to existence; this permitted him to envisage a form of spiritual and moral progress that would not involve eugenic tampering. It also allowed him to escape the reductionism of pure Darwinism in relation to cultural achievement and aesthetics since they also could be attributed to our higher, spiritual nature. Wallace's thought, in fact, spans both sides of the contemporary debate within science and also constitutes something of a compromise position where the related dispute between religion and science is concerned. Gould, however, as an atheistic evolutionist, has had no other option but to look to cultural evolution alone in order to explain such phenomena as well as for a solution to the question of human moral progress.

---

[36] Stephen Jay Gould, *The Structure of Evolutionary Theory* (Cambridge, MA: The Belknap Press of Harvard University Press, 2002).

Before we begin to address the question of which side in this ongoing and bitterly contested dispute is the most plausible on the subject of human behavior and the nature of human progress, a brief word must be said about the influence of Lamarck on Teilhard de Chardin and hence on some contemporary theological speculation. De Chardin incorporated Lamarck's notion of orthogenesis into his theology of evolution, depicting salvation history as an ascent toward moral perfection and the manifestation of Christ in the Parousia, or second coming. Within this ascent, de Chardin views evil, both physical and moral, as a necessary byproduct of evolutionary processes.[37] In my opinion, the latter viewpoint neutralizes the theological problem of evil to an unacceptable degree and, as we are about to see, is not supported by either the scientific evidence or the evidence of history.

Darwinism versus Lamarckism: Natural Selection or Punctuated Equilibrium?

With regard to scientific credibility, it is generally accepted in scientific circles that natural selection is the most important mechanism of evolution; punctuated equilibrium is considered to have a limited evolutionary role at best. This acceptance accrues from the fact that natural selection is one of the oldest unfalsified theories in the history of science. It has demonstrated great explanatory power and has been a unifying force throughout every field of biology; it has also withstood and absorbed all

---

[37] See "Comment je vois," an unpublished essay by de Chardin, cited by Piet Schoonenberg in *Man and Sin* (London: Sheed and Ward, 1965), pp. 43–44.

scientific challenges over the past century.[38] The theory of evolution by natural selection, in fact, goes from strength to strength having recently been reinforced by the evidence of the chromosomes; changes in gene sequences on chromosomes occur through natural selection. Through interpreting these changes, geneticists have proven the fact of evolution and traced the route taken by our remote ancestors from Africa, from where they spread out to populate the world. It is now known that a mere 2 percent of the genome separates the human species from chimpanzees; in the future, it is claimed, geneticists will be able to explain the role of genes not just in the anatomical differences between humans and chimpanzees but also in the behavioral differences. In the meantime, the discipline of behavioral genetics, though new, has shown that many aspects of mental, psychological, and personal development are at least partly heritable. The new field of evolutionary psychology (the latest incarnation of sociobiology) has also had significant success in showing that the brain is an evolved organ and has amassed a growing body of evidence in support of the claim that many aspects of human behavior are universal and can be explained in adaptive terms. It is the latter evolutionary discipline, in conjunction with neuroscience and analysis of the genome that will, over time, develop the capacity to test out Wallace's hypotheses regarding what natural selection can and cannot achieve with regard to human intellectual and moral capabilities.

Yet the question of the impact of evolution on human behavior remains fiercely disputed within science; this is due

---

[38] For a comprehensive discussion of the debate between Dawkins and Gould, the best source I have come across is the work of Ruse, especially *The Darwinian Paradigm.*

in the main to the potential application of future genetic technologies for eugenic purposes. In the very near future we, as a species, may be able to influence or even take control of our own evolution; genetic technologies will give us the power to manipulate the building blocks of human nature, along with the potential to achieve great good or immense harm in relation to human well-being. If it turns out that neo-Darwinist theory is correct, and that genes for behavior really are proportioned in such a way that future moral progress is highly problematic, is it not inevitable that the scientific focus will turn to morality as the most important form of progress, the one on which all other forms depend? Is it not also highly likely that possibilities for behavioral "improvement" through prosthetic means or genetic manipulation will overcome present inhibitions? There is also a danger that a fatalistic acceptance of the inevitability of dominant elites will take hold, and, further, that such elites in the fields of science, medicine, and politics will have the power to decide how the various genetic technologies should be used in the improvement not only of human physical and mental health but also of human behavior. This would mean that definitions of what constitutes "good" and "evil" would be formulated by the powerful, and such a scenario would almost certainly lead to abuse of the weak by the strong. In his excellent book *The Political Gene*,[39] Dennis Sewell chronicles the extent of the impact that the "might is right" strand of Darwinism has had on political policy in the West up to now, particularly in relation to a popular movement known as social Darwinism that emerged in the late nineteenth century and that advocated the applica-

---

[39] Dennis Sewell, *The Political Gene: How Darwin's Ideas Changed Politics* (London: Picador, 2009).

tion of the theory of natural selection to society. Its creator, the philosopher Herbert Spencer, argued that no social support should be given to the poorer, weaker members of society in the hope that their "weak" characteristics would be eliminated from the gene pool. Sewell makes all too clear the likelihood of new genetic technologies being disastrously incorporated into political theory and policy in the near future in a similar response to neo-Darwinism. As a matter of interest, it must be pointed out here that Darwin himself was firmly opposed to social Darwinism; its ideology was based on a conflation of his work with that of Spencer, though it owed far more to the latter than to the former. Darwin, in fact, considered the human capacity for empathy and compassion with our fellow man to be the "noblest part" of our nature, the basis of all civilization and all potential for future moral progress.[40] There is a danger however that this second wave of Darwinian theory, driven as it is by a new understanding of the evolution of cooperative behaviors, could stigmatize those who are shy, socially inept, or unable to project a convincing image as a social cooperator. Indeed, evolutionary biologists Thornhill and Palmer have argued that evolution confers greater benefits on those who can use their rationality to dissemble altruism than on those who actually are altruistic:

> Humans are social strategists par excellence [and] have evolved to present themselves as altruists. . . . This presentation involves "indirect reciprocity," in which the displayer obtains a reputation or a social image as a potential cooperator. Here the word "indirect"

[40] Darwin, *Descent,* Vol. 1, pp. 168–169.

> pertains to the fact that the returns to the altruistic individual come from onlookers rather than from the individual helped. Altruistic display, in which one attempts to portray oneself as concerned about anonymous others and about the human species in general, is crucial to an individual's development of a reputation as a social ally.[41]

The question now facing both the sciences and the humanities (which, of course, includes religion) is simply this: Is progress possible, and if so, how is it to be defined? With this question in mind, I would suggest that the technologically fueled wars and catastrophes of the twentieth century provide us with a unique vantage point from which to assess the Enlightenment project of an inevitable, upward, progressive climb for humanity. The Enlightenment itself was, of course—for all of its emphasis on the power of human rationality—the product of physical as well as intellectual revolution, particularly in France and North America. French thinker Ernest Renan wrote in 1882 of the "deeds of violence that have taken place at the commencement of all political formations" and of how "Unity is ever achieved by brutality."[42] In the context of the evolutionary claim that there are enough disinterestedly altruistic individuals in societies to effect progress through peaceful means, the broad sweep of history would seem to indicate that "progress" comes about mainly through the efforts of a minority of individuals who

---

[41] Randy Thornhill and Craig T. Palmer, *Rape: A Natural History* (Cambridge, MA: The MIT Press, 2000), p. 116.
[42] From his famous essay "What is a Nation," first delivered at the Sorbonne in 1882.

band together to seize power with or without popular support. Can this recurring pattern be described as progress in the sense intended by Enlightenment thinkers such as Marx, or is it evidence for Nietzsche's "eternal recurrence"?

Many leading contemporary intellectuals from various spheres have commented on the implications of technological progress for Enlightenment optimism concerning human rationality. Kenan Malik, a social commentator and philosopher of science points out that

> The barbarous history of the 20th Century—two world wars, the Holocaust and ethnic cleansing—has left many people disillusioned about what it means to be human. Every impression that man makes upon the world seems to be for the worse. The attempt to master nature has led to global warming and species depletion. The attempt to master society has led to Auschwitz and the gulags. The result has been a growth of anti-humanism, of despair about human capacities, a view of human reason as a force for destruction rather than betterment.[43]

Is it not, in fact, the case—as Freud claimed—that technological progress seems to militate against moral progress, since it is all too often used by "the fittest" in their drive to achieve power, status, and wealth? In this way reason is put at the service of Nietzsche's instinctive "will to power." John Gray, successor

---

[43] Kenan Malik, "Man, the Utterly Exceptional Beast," printed in *The Sunday Times*, 22 October, 2000. Excerpted from his book *Man, Beast and Zombie* (London: Weidenfeld and Nicolson, 2000).

to Nietzsche, and a leading contemporary philosopher of Darwinism, remarks on the irony inherent in the fact that atheistic evolutionary biologists—often militant opponents of Christianity—are in fact promoting a Christian worldview in their insistence that humans, uniquely among all earthly species, can defy the laws of their nature and progress:

> To believe in progress is to believe that, by using the new powers given us by growing scientific knowledge, humans can free themselves from the limits that frame the lives of other animals. This is the hope of nearly everyone nowadays, but it is groundless. For though human knowledge will very likely continue to grow and with it human power, the human animal will stay the same: a highly inventive species that is also one of the most predatory and destructive.[44]

Gray argues that human technological inventiveness is as "natural as a spider's web" and as impossible for us to control as it would be for a spider to weave anything other than a web; the pace and type of technological development, he asserts, is part of the inevitable struggle to survive and therefore determines us rather than the other way around. Thus, the hopes of neo-Darwinists such as E. O. Wilson that genetic engineering will soon give humanity godlike powers "to take control of its own ultimate fate," or of Richard Dawkins that we can defy our "selfish genes" now that we know about them with their cultural equivalent "memes" are, he suggests, deluded. As he points

[44] John Gray, *Straw Dogs: Thoughts on Human and Other Animals* (London: Granta, 2002), p. 4.

out, "Pogroms are as old as Christendom; but without railways, the telegraph and poison gas there would have been no Holocaust. There have always been tyrannies; but without modern means of transport and communication, Stalin and Mao could not have built their gulags. Humanity's worst crimes were made possible only by modern technology." However, once something becomes technologically possible it will be attempted, and Gray believes it likely that at some stage during the present century human nature will be "scientifically remodeled" and

> that it will be done in a haphazard way as an upshot of struggles in the murky realm where big business, organized crime, and the hidden parts of government vie for control. If the human species is re-engineered it will not be the result of humanity assuming a god-like control of its destiny. It will be another twist of man's fate.[45]

In *Homo Deus: A Brief History of Tomorrow*, the Israeli historian Yuval Harari, who interprets history through a Darwinian lens, focuses on the growing power of science and his conviction that it will lead to a new world in which "the gap between those who get on board and those left behind will be larger than the gap between industrial empires and agrarian tribes, larger even than the gap between *Sapiens* and Neanderthals." This, he argues, "is the next stage of evolution. This is Homo Deus." Harari even goes so far as to speculate that, due to the creation of artificial intelligence (AI) a dystopian situation could emerge

---

[45] Ibid., pp. 5–16.

in which human find themselves under the control of super-intelligent robotic beings.[46]

In *Black Mass*, Gray makes an excellent case in support of his thesis that theories of progress depend on a teleological or purposeful view of history that has its roots in Christianity and was coopted into the secular vision of the Enlightenment; to maintain such a view, he maintains—like Russell—that belief in the supernatural would be necessary.[47] At this point we may also take note of the fact that it is the Darwinian strand of evolutionary philosophy that has persisted most successfully into the twentieth and twenty-first centuries, and this is a further indication that natural selection is indeed the main mechanism of evolution.

A poignant judgement from the field of literature on the twentieth century's terrible implications for Enlightenment optimism is made in Bernhard Schlink's *The Reader*. Its main character, a legal historian in the postwar period who has been traumatized by his relationship with a concentration camp guard, muses that the legal codes of the Enlightenment were

> based on the belief that a good order is intrinsic to the world and that therefore the world can be brought into good order. To see how legal provisions were created paragraph by paragraph out of this belief as solemn guardians of this good order, and worked into laws that strove for beauty and by their very beauty for truth, made me happy. For a long time I believed

---

[46] Harari, *Homo Deus* (New York: Harper, 2017), front and back covers.

[47] John Gray, *Black Mass: Apocalyptic Religion and the Death of Utopia* (London: Penguin, 2007), p. 41.

> that there was progress in the history of law, a development towards greater beauty and truth, rationality and humanity, despite terrible setbacks and retreats. Once it became clear to me that this belief was a chimera, I began playing with a different image of the course of legal history. In this one it still has a purpose, but the goal it finally attains, after countless disruptions, confusions, and delusions, is the beginning, its own original starting point, which once reached, must be set off from again.[48]

What, one wonders, would the great intellectuals and politicians of the Enlightenment period—men such as Locke, Bentham, Voltaire, Diderot, Condorcet, and Franklin—have to say about human reason and its possibilities in the aftermath of the twentieth century? Would their optimism remain undimmed? I doubt it.

The evidence of recent history would certainly seem to support a Darwinian-Nietzschean view of human nature rather than that of the Enlightenment and its neo-Lamarckist derivative. The pendulum has swung, over the course of the last two centuries, from unbridled optimism to growing pessimism where human nature is concerned. Yet belief in progress is still very much alive in Western society, which remains the brainchild of the Enlightenment in law, politics, the social sciences, and the humanities. This is why the idea that genetic technologies can bring about improvement in the human species is at once so dangerous and so attractive, and why the contemporary dispute raging within scientific circles is of such crucial importance. At this point, let

[48] Bernhard Schlink, *The Reader* (London: Phoenix, 1995), p. 179.

us remind ourselves of the relevance of the intrascientific debate to religion. Firstly, both perspectives on evolution and human behavior pose a challenge to theistic religion generally. The neo-Lamarckian view that humanity has the power to perfect its own nature does away with the need for salvation preached by Christianity in particular (though all three of the Abrahamic faiths portray a human nature in need of supernatural help in order to reach fulfillment); meanwhile, neo-Darwinism undermines the view that God is all good. If true, it casts serious doubt on the existence of the omniscient, omnipotent, benign deity it—along with Judaism and Islam—portrays. I believe that until the dual challenge posed by evolution to religion is resolved in a way that cannot be easily dismissed by skeptical scientists, there can be no meaningful reconciliation between evolutionary biology and religion. Even if, for example, the intelligent design viewpoint were to be accepted as a valid one by the scientific community in the foreseeable future—unlikely as such a scenario might seem—the challenge posed by the evolution of evil would remain unanswered. As Darwin himself pointed out, if God is responsible for designing human morality either directly or indirectly, then God is responsible for human evil; if, on the other hand, human beings had free will and chose a morally disadvantageous evolutionary path, why did God not intervene to prevent its worst effects? Logically—as Darwin also pointed out—intelligent design would be more important in the moral sphere than in any other aspect of the evolutionary process. These are questions, I believe, that cannot be shirked; not everyone is blessed with an unshakeable faith, nor should unshakeable faith permit the burying of one's head in the sand where compelling scientific evidence is concerned. Atheists, agnostics, and skeptics of all kinds tend to be critical thinkers who require a degree of reason as well as faith on the spiritual quest that we must all make in one way or another; they are entitled to ask questions and have their legitimate concerns addressed. In addition, if we are to avoid excessive scientism in public debate, religion must be able to deal with the arguments of scientists; the ethics of the future use or abuse of genetic technologies, and hence the whole question of progress, is highly relevant to religion, though the religious viewpoint is largely excluded from consideration due to the growing impact of scientism.

It is my opinion that the reemergence of the problem of evil in the guise of genetic determinism is the most serious blow that science has dealt to theistic religion and has brought evolutionary biology itself to a dangerous impasse. The onus, however, is on religion to grasp the nettle that has been identified by science. Needless to say, there can be no return to a prescientific or preevolutionary mentality, and I believe that the insights of evolutionary biology are as relevant to religion as religion is necessary to bring the insights of the former discipline to full fruition. Since the scientific, historical, and philosophical data support the neo-Darwinist rather than the neo-Lamarckist view of progress, in the following chapter we will confront this new challenge to theistic religion by undertaking further analysis of the behavioral issues raised by the operation of ESSs: Was their evolution inevitable? This is the vital question, since the inevitable evolution of such strategies as lion and lamb, chimpanzee, bonobo, and elephant would, in my opinion, pronounce the death knell for theism. Interestingly, the similarity between the Darwinian account of human nature and the Christian one of a humanity damaged by an "original sin" has been remarked on by leading evolutionary commentators such as E. O. Wilson and the British philosopher of science Michael Ruse. The appropriation of Darwinism into Christianity, however, would require the cause of evolutionary evil—natural selection—to be attributed to the deity; so while the existence of evil is explained, and the characteristics that cause it are more or less identical to those described by Christianity, the problem of evil is not resolved either for Christianity or the other theistic religions. In the following chapter, therefore, we will subject the scientific evidence to rigorous analysis to determine whether the evolutionary process had to take the course that it did. Only when that question is decided will it be possible to properly evaluate the claims of both neo-Darwinism and neo-Lamarckism from a theistic perspective.

Chapter Three
Searching for a New Theodicy: Before Evil Evolved

Before we begin our analysis of the scientific evidence, I would like to say a few words about methodology. There are various views in the West about the relationship between science and religion, two of them at extreme ends of the spectrum. At the scientific end is the materialist view that matter can explain everything and that science is the highest form of knowledge; at the religious end there is a corresponding refusal to engage with science at all on the basis that faith takes precedence over all other forms of knowledge. Two somewhat less extreme viewpoints are the independence model and the conflict model. The first holds that science and religion should be kept separate as two fields of knowledge, since they deal with different aspects of human experience; in this view, religion answers "why" questions, while science answers "how" questions. This is to ignore, however, the many and fruitful ways in which religion and science can and have interacted and overlapped. After the initial theological shock of Galileo's heliocentrism, for example, it became accepted in mainstream theology that the Bible need not be taken literally. The Genesis creation accounts were understood to have significant metaphorical and mythological dimensions that have greatly enriched interpretation of the texts. In addition, it was scientific methodology that gave rise to the historical-critical method of biblical interpretation, a theological advance that has revolutionized the discipline of biblical theology and enabled a historical understanding of the New Testament in particular that sheds greater light on the theological significance of the Gospels. The independence model also ignores the fact that culturally our beliefs about what constitutes reality are very much shaped by the practical achievements of science, and this inevitably impacts our religious beliefs to a greater or lesser degree.

The second approach is the view that although science and religion often appear to be in conflict, this conflict is only apparent and that the passage of time will resolve any contradictions and show the two fields to be in harmony. The latter viewpoint encourages engagement between the two fields. In this regard there can be no doubt that Christian theology in particular has always had a fascination with evolution, and this is undoubtedly due to the strong focus of Judeo-Christianity on a creator God and on the essential goodness of creation. Despite the fact that the greatest challenge posed by science to Christianity is certainly that of Darwinian evolution to the doctrines of creation and salvation, the passage of time has led to the development of a perspective on human nature in Darwinism that is almost identical to that of Christianity. The latter, recent consilience between religion and science is all the more striking because it breaks down the distinction asserted by the independence model; evolutionary biology explains not only how we evolved biologically and morally but goes a long way toward explaining *why* we behave as we do. The only stumbling block to full harmonization is, as we saw in the previous chapter, the challenge of Darwinism to the Christian concept of an all-good, all-powerful God. I would argue that when a conflict between the two spheres that is an obstacle to the tantalizing potential for synthesis emerges, it would amount to theological negligence not to interrogate the scientific evidence to seek out a resolution to the problem. As well as a theological imperative, there is also an important philosophical one. The nature of the epistemological relationship between science and religion remains unresolved and highly controversial; a resolution of one of the most serious challenges posed by evolutionary biology to Christianity would have the immense bonus of illuminating the relationship between scientific truth and religious truth overall.

It is of course the perspective of faith that is the catalyst for questioning the findings of evolutionary biology and that will determine the questions to be asked. However, the science that needs to be interrogated must be interrogated strictly as science. The bottom line is that if both Christianity and Darwinism are true, then logically the evolutionary outcome cannot have been inevitable. The latter conclusion will of course guide the questions that are asked of the empirical evidence, the most important of which has to be: Does the empirical evidence indicate that the evolution of intractable evil was inevitable? Keeping all of this in mind, we will now turn to an examination of what evolutionary biology and its related disciplines have to tell us about the primal events of our prehistory.

Natural Selection and the Struggle for Existence

Natural selection comes into operation when there is a "struggle for existence" that forces the individuals of a species to compete for resources. For thousands of millions of years before the arrival of *Homo sapiens* on Earth, natural selection had been a competitive but unconscious process; in the case of animals, an unconscious and instinctive process. However, with the emergence of the conscious and rational human species, evolutionary competition would also have become conscious and rational, pitting man against man and all men against other species and the environment. In *The Descent of Man*, Darwin points out that the evolution of morality would have been dependent on the acquirement of a high level of intellectual ability. He defines a moral being as "one who is capable of comparing his past and future actions or motives, and of approving or disapproving of them."[49] By the time moral choice became a possibility for the human species, a high degree of intellectual ability would already characterize it. It seems clear that the same rational powers that endowed the first humans with consciousness also endowed them with free will; for the first time in the billions of years since the emergence of life on earth, moral decision making had become a possibility. The Darwinian challenge to theism becomes clear if, as Darwin and many of his successors believe, the combination of natural selection with human rationality *inevitably* leads to the levels of moral evil asserted by Darwinism and neo-Darwinism. If God is the creator of natural selection and, by implication, human nature, this would imply that God is the author of evil. The solution to this theological conundrum lies, I believe, in finding out whether the choices made by the first humans really were inevitable.

We are what is known in evolutionary biology as a diploid social species, a term used to describe species whose individuals share at most 50 percent of their genes; in such species, short-term gain can militate against the long-term welfare of the group. When, for example, clonal organisms—which share 100 percent

---

[49] Darwin, *Descent*, Vol. 1, pp. 72–73, p. 88.

of their genes—form colonies, selection ignores the separate organisms and selects the colonies, which compete vigorously as groups; individual clones are identical, so individual competition is pointless. Clonal aphids, for example, when infested with lethal parasitic wasps, altruistically commit suicide rather than staying in their colony to infect their 100 percent genetically related counterparts. In contrast, diploid organisms, like ourselves, vary and compete, undermining group cohesion.[50] In diploid species driven by instinct alone, selection operating at the level of the individual is compatible with ecological balance and harmony; this is not the case with humans, however, who are able to foresee long-term contingencies and employ their powers of reason selfishly and cruelly in the service of their basest instincts in ways that are harmful to the group, the species—and the planet—if they so choose. Humans, alone in the natural world, however, also have the ability to work and plan for the long-term welfare of the group. Precisely because of its intellectual ability to balance individual needs against those of the group, the human species is the only diploid species in which group selection for altruism, as envisaged by neo-Lamarckists such as Gould, might have occurred. As I will argue in the following chapters, however, the latter type of moral evolution not only could have occurred, but should have occurred.

A certain degree of aggression, selfishness, and competitiveness would have had to evolve and become part of human psychology for the species to survive, thrive, and, from a theological perspective, for virtue to be possible; it is the degree of evil and its stark consequences that present a problem to the

[50] See Patricia A. Williams, *Doing without Adam and Eve* (Minneapolis: Fortress Press, 2001), p. 129.

concept of a good God. From the perspective of faith, we will now hypothesize that human beings must have had as much freedom of the will as would have been necessary to resist the evolution of evil to the level that it has actually evolved. Therefore, again from the perspective of faith, we can speculate that evolution did not have to take the course that it did. A moral choice or choices were made that caused it to take that particular direction, and the similarities between the Darwinian and Christian accounts of human nature are no coincidence. Leading Anglican theologian and evolutionary theist Keith Ward has remarked that

> In the course of evolution, there must have been a first moment of conscious moral choice. That is the point at which the "fall of humanity" began and humans were estranged from that natural fellowship with God which should have been theirs, and from their natural ability to relate unselfishly to one another.[51]

I am, of course, in full agreement with Ward on this point. In light of the array of opinion to the contrary however, I believe that the argument for free will at the outset of human history must be subjected to the scientific evidence. A second important reason for exploring this question in detail is to find out if science can expand our theological knowledge of the impact of sin both at the individual level and at the level of the species. The classical doctrine focuses on the impact of sin on the individual; we now know, however, that as a social species, human

---

[51] Keith Ward, *God, Faith and the New Millennium* (Oxford: Oneworld, 1998), p. 42.

behavior evolved in the interactions of individuals not just with one another, but with the various groupings to which each individual belongs, the largest of which is society. From this it can be inferred that if indeed the first humans did have free will, their transgression would have affected not only the moral evolution of individuals but that of the species itself.

We now begin an exploration of the scientific data to find out if it permits the possibility that evolution might have taken a different course or courses to the one that emerged from that primordial "struggle for existence" at the outset of human history. We begin by considering Darwin's speculations concerning the earliest possible manifestations of the moral sense.

From Ancestral Humans to Primitive Tribes

In trying to visualize the behavior of the first humans, Darwin drew on the experiences he had of primitive tribespeople in South America, Australia, and New Zealand during the course of his *Beagle* voyage. His very first sightings of a "savage" tribe in Tierra del Fuego set the tone for his future speculations. The idealistic captain of the *Beagle*, Robert Fitzroy, had hoped to set up an Anglican mission there, and as they sailed along its coastline Darwin was stunned by what he saw:

> I shall never forget how savage and wild one group was. —Four or five men suddenly appeared on a cliff near to us,—they were absolutely naked and with long streaming hair; springing from the ground and waving their arms around their heads, they sent forth most hideous yells. Their appearance was

> so strange, that it was scarcely like that of earthly inhabitants.[52]

The impact of this first sighting never left him and colored ever afterward his views on the evolution of humanity that, as a result, were strikingly different to those of Wallace; he would always draw a sharp distinction between the very earliest humans and those who would have emerged as natural selection took effect on their behavior. The wild men of Tierra del Fuego represented to him the inevitable impact of natural selection on human rational and moral powers. Darwin was appalled by the cruelty of primitive tribesmen and women, their delight in torture, the promiscuity and licentiousness particularly of the men who treated their wives as slaves; he was equally horrified at the casual attitude of the women toward infanticide. Nor—despite his hopes for moral progress—was he convinced that "civilization" did anything more than coat such barbarism with a thin veneer. Darwin's travels and his experiences of various native tribes convinced him that this veneer could quickly be stripped away by circumstances; he noted how he himself and his *Beagle* friends had toughened up and hardened under primitive conditions. The ability to degenerate rapidly from a well-mannered European "into a hardened slave-master or berry-picking refugee" was something that both he and Fitzroy had observed. He "was forced to acknowledge that the gauzy film of culture was nothing but an outer garment for humanity, acquired or lost in response to the individual milieu"[53] and that no group of human

---

[52] Cited in Browne, *Voyaging*, p. 240.
[53] Ibid., pp. 247–248; p. 250.

beings, from urbane and sophisticated Englishmen to the wildest savages, was exempt from the most basic of impulses.

The very first humans, he felt, would have been markedly gentler and less aggressive than the "savages" he had encountered on his travels. Primeval man, Darwin surmised, when he had only barely attained the rank of humanity, would have been governed more by his instincts and less by reason than either indigenous tribesmen and women or ourselves. The first humans would also have been far more sexually continent than their descendants, since they would at first have been under the sway of their habitual instincts. Thus, there would have been no promiscuity and no adultery. A man would have been loyal to his wife or, in some few cases perhaps, wives, defending them from all kinds of enemies and providing food for them and their children. They would not yet have developed the ability to look forward to "distant contingencies" and would have been incapable of serious deception or treachery. Due to the strength of the familial instincts, the first humans "would not . . . have partially lost one of the strongest of all instincts, common to all the lower animals, namely the love of their young offspring; and consequently they would not have practiced infanticide."[54] The inability to plot and plan for the future would have preserved for a time the affectionate social instincts that species of apes traditionally display. Initially, there would have been no premeditated violence or treachery in the pursuit of women, wealth, or power. Why didn't this happier state of affairs persist?

It must be kept in mind, of course, that the first humans, though undoubtedly a gentler and more affectionate species than ourselves, would have had competitive and aggressive

---

[54] Darwin, *Descent,* Vol. 2, pp. 367–368, and frequently.

instincts and impulses as well as affectionate and cooperative ones. What they would not yet have developed was the ability to plot and plan for the short-term and long-term gratification of their own individual interests against the interests of the group. At that earliest stage of human existence, as soon as the powers of reason had become sufficiently developed, however, the moral faculties of the first humans would have come under the scrutiny of natural selection; in neo-Darwinist terms, there would have been a period of evolutionary instability as previous behavioral strategies became unnecessary or redundant. The earliest humans would then, morally speaking, have had two options: they could come under the sway of new evolutionarily stable strategies, or dispense altogether with the need for any such intervention on the part of nature. Just as nature, at a certain point in the evolution of human intelligence, had become redundant with regard to human biological adaptation, at a later time its role in the evolution of moral behavior could have ceased, and a neo-Lamarckian type of cultural moral evolution would have occurred. We analyzed in chapter 2 the composition of "evolutionarily stable strategies" that balance aggression and cooperation at stable rates, maintain the status quo, and are defined by complex relationships. To change these strategies at our own stage of evolution would appear to be highly problematic without some sort of artificial intervention.

The salient point in relation to human moral evolution is that with the acquisition of consciousness and rationality our earliest ancestors, alone among all other earthly species, had a choice as to how they would respond to the various situations confronting them. The same rational powers that enabled them to devise individual survival strategies at the expense of increased cooperation would also have allowed them to put the welfare of the tribe above their own individual welfare, when to do so would have been to the overall benefit of the tribe. At the earliest stage of human history, individuals would have lived in small troops or tribes and seen firsthand on a daily basis the impact of selfish, cruel, or treacherous behaviors on their fellow tribesmen and women and on the group as a whole. Logically, it would have been far easier in such circumstances to grasp the importance of cooperation; what actually seems to have happened, however, was a descent into anarchy. This has been described by philosopher of science Daniel Dennett as "a grim default position of universal treachery"[55] and—somewhat earlier—by Thomas Hobbes as "a war of all against all." This state would have been followed by the development of new evolutionarily stable strategies, of the kind described by Richard Dawkins, and characterized the period of evolution that produced the primitive races described by Darwin and Wallace. There then ensued a process of civilization rather than further moral evolution, resulting in the so-called civilized races.

Keeping all of this in mind, and with the help of Darwin's speculations on the evolution of conscience, we can now begin to ponder the question of what actually happened and what *could* have happened at the outset of human history.

The Evolution of Conscience

In trying to envisage human behavior during the earliest period of human existence, Darwin's speculations on how conscience evolved are of the utmost importance. Ironically, in his minutely detailed and painstaking analysis, Darwin's preoccupations were with the religious controversies stirred up by the *Origin* in a

[55] Daniel Dennett, *Freedom Evolves* (New York: Penguin, 2003), p. 149.

society that was, in the main, still unquestioningly devout. His interest is not so much in conscience per se as in demonstrating his conviction that there is no clear distinction between human intellectual and moral faculties and those of animals that would imply the separate creation of humanity. He also wants to show that there is a naturalistic explanation for the existence of conscience that dispenses with the need to look for supernatural input.

Darwin precedes his discussion of moral evolution by tracing the evolution of human intelligence, since he believed highly developed intellectual faculties to be a prerequisite for the development of conscience. He looks first of all at the role of animal emotions, especially complex or what he terms "intellectual emotions," in the development of the "higher mental powers" of the primates that would eventually become human. Citing numerous examples of the behavior of various animals, he shows that man and the higher animals possess "similar passions, affections, and emotions, even the more complex ones; they feel wonder and curiosity; they possess the same faculties of imitation, attention, memory, imagination, and reason, though in very different degrees." He describes, for example, the dread that monkeys have of snakes; yet their curiosity regarding them is so great that they have been known to satisfy it "in a most human fashion" by lifting up the lid of a box in which snakes were kept to peek in. The principle of imitation, he says, is a marked characteristic of man, as it is of monkeys, "who are well-known to be ridiculous mockers." Darwin points out the undeniable fact that animals often have excellent memories for people and places. While admitting that imagination is one of the highest faculties in man, he points out that dogs, cats, horses, and probably all the higher animals are known to have vivid dreams as is shown by their movements while sleeping, and this indicates that they possess some power of imagination. Darwin places reason at the summit of human faculties, yet "few persons any longer dispute that animals possess some power of reasoning. Animals may constantly be seen to pause, deliberate, and resolve." He gives several examples, including that of some American monkeys that, on initially being given eggs, smashed them and lost most of their contents. The next time they treated the eggs as fragile objects, picking off bits of shell with their fingers. He also describes the retriever of a friend that tried to carry two ducks over a stream, but couldn't manage both. The dog, which had never been known to so much as ruffle the feather of a bird, deliberately killed one and brought over the other, returning immediately for the dead bird. Darwin remarks that the chimpanzee in a state of nature cracks nuts with a stone and recalls a young orangutan that he saw "put a stick into a crevice, slip his hand to the other end, and use it in the proper manner as a lever."[56]

Darwin places the faculty of language in humans as the most important of all the faculties in the evolution of human intelligence and morality, remarking that the latter has "justly been considered as one of the chief distinctions between man and the lower animals." Darwin qualifies this however, referring to his own observations and numerous authorities in describing the various vocal sounds of birds, dogs, and apes in particular. Monkeys, when excited, make at least six distinct sounds which in turn excite other monkeys; the facial expressions and gestures of monkeys are understood by us while they, in turn, partly understand ours. Darwin also refers to the interesting phenomenon

---

[56] *Descent*, Vol. 1, pp. 45–48, 51–52.

of dogs that, once domesticated, learn to bark in at least four different tones. There is the bark of eagerness in the chase or while playing, the angry bark, the yelping or howling bark of despair, and the bark of joy when being brought out for a walk. He concludes that it is not language in itself that is unique to humans, but the power of *articulate* language:

> Articulate language is, however, peculiar to man; but he uses in common with the lower animals inarticulate cries to express his meaning, aided by gestures and the movements of the muscles of the face. This especially holds good with the more simple and vivid feelings, which are but little connected with our higher intelligence. Our cries of pain, fear, surprise, anger, together with their appropriate actions and the murmur of a mother to her beloved child, are more expressive than any words. It is not [even] the mere power of articulation that distinguishes man from other animals, for as everyone knows, parrots can talk; but it is his large power of connecting definite sounds with definite ideas; and this obviously depends on the development of his mental faculties.[57]

The difference between man and animals is only one of degree, according to Darwin. The question remains, however, as to how man developed his far superior intelligence. Darwin believes that the evolution of human intelligence would have begun with the ability to articulate:

[57] Ibid., Vol. 1, pp. 53–55.

> I cannot doubt that language owes its origin to the imitation and modification, aided by signs and gestures, of various natural sounds, the voices of other animals, and man's own instinctive cries. . . . As monkeys certainly understand much that is said to them by man, and as in a state of nature they utter signal-cries of danger to their fellows, it does not appear altogether incredible, that some unusually wise ape-like animal should have thought of imitating the growl of a beast of prey, so as to indicate to his fellow monkeys the nature of the expected danger. And this would have been a first step in the formation of language.[58]

Those "wise ape-like animals" that were able to thus warn one another would obviously have prospered and spread their genes. The vocal organs would therefore have evolved through succeeding generations, since the power of communication through language would, of course, confer great advantage on any species that possessed it. Most significantly, according to Darwin, this initially rudimentary power of speech as it evolved and developed would have facilitated the power to think, imagine, and speculate.

With his analysis of the evolution of language, Darwin lays the groundwork for his subsequent discussion of the human moral sense; for with the power of articulate language would come self-consciousness, individuality, the power of abstraction, and the ability to formulate general ideas. Such faculties, he argues, would be a prior necessity for the development of conscience and the emergence of a moral sense.[59]

---

[58] Ibid., pp. 56–57.
[59] Ibid., pp. 57–62.

The Role of the Social Instincts in the Development of Conscience

Apart from a high level of intelligence, Darwin specifies another quality of the human species essential to the development of a moral sense, though it would have preceded the evolution of the intellectual faculties. Humanity, he points out, is a social species and "any animal whatever, endowed with well-marked social instincts, would inevitably acquire a moral sense or conscience, as soon as its intellectual powers had become as well developed, or nearly as well developed, as in man." This is because social animals are distinguished by their love and empathy for each other. They take pleasure in the company of others, and are eager to aid their fellows in various ways.[60] Social animals, according to Darwin, were rendered so by natural selection, which caused in them the evolution of feelings of affection for one another and sensations of pleasure in each other's company: "These sensations were first developed, in order that those animals which would profit by living in society, should be induced to live together. . . . The feeling of pleasure from society is probably an extension of the parental or filial affections [kin selection]." As usual in making his point, Darwin provides many fascinating and amusing examples of how social instincts operate in the animal world. He mentions the hamadryas baboons that turn over stones to find insects; when they come upon a particularly large one, they all stand around to turn it over and share the spoils. Social animals also mutually defend one another, and to illustrate this Darwin tells the story of a troop of baboons under attack by dogs. They all climbed a nearby mountain to escape except for a young one of six months

---

[60] Ibid., pp. 71–72; p. 80.

old that was left behind. The young baboon climbed onto a rock where he was surrounded by the pack of dogs and began to call loudly for aid. One of the largest males of the troop, "a true hero" according to Darwin, came back down, slowly approached the young one so as not to frighten him, and then swiftly scooped him up, making his escape before the dogs could attack. Another of Darwin's favorite anecdotes emerged from his frequent visits to the London zoo and concerns a "sympathetic and heroic" American monkey who lived there. The monkey had established a warm friendship with his keeper, who on one occasion showed Darwin some deep wounds on the nape of his neck that had been inflicted on him by a fierce baboon while he was kneeling on the floor of his cage. The little American monkey, which lived in the same cage, was normally terrified of the big baboon; nonetheless when he saw that his friend the keeper was in danger he rushed over, screaming and biting at the baboon in order to distract him. The keeper was then able to escape, saving his life. Darwin's admiration for the strong ties of affection, bravery, and solidarity he saw in social species that are closely related to us was undoubtedly a factor in his conviction that the very first humans would have been a far kinder and gentler species than the Fuegians or us, their "civilized" counterparts. At the close of *The Descent of Man* he remarks:

> He who has seen a savage in his native land will not feel much shame, if forced to acknowledge that the blood of some more humble creature flows in his veins. For my own part I would as soon be descended from that heroic little monkey, who braved his dreaded enemy in order to save the life of his keeper; or from that old baboon, who, descending from the mountains,

> carried away in triumph his young comrade from a crowd of astonished dogs—as from a savage who delights to torture his enemies, offers up bloody sacrifices, practices infanticide without remorse, treats his wives like slaves, knows no decency, and is haunted by the grossest superstitions.[61]

Darwin repeatedly stresses his opinion that in social species, the social instinct is the most enduring and persistent, if not the most powerful of the instincts:

> Now with those animals which live permanently in a body, the social instincts are ever present and persistent. Such animals are always ready to utter the danger-signal, to defend the community, and to give aid to their fellows in accordance with their habits; they feel at all times, without the stimulus of any special passion or desire, some degree of love and sympathy for them; they are unhappy if long separated from them, and always happy to be in their company.[62]

He therefore argues that as soon as the mental faculties had become sufficiently highly developed to allow consciousness of past actions and motives, the satisfaction of some temporarily powerful, selfish instinct would, when assuaged, eventually give way to feelings of guilt once the ever-present social instinct re-asserted itself. "It is clear," says Darwin, "that many instinctive desires, such as hunger, are in their nature of short duration; and after being satisfied are

---

[61] Darwin, *Descent*, Vol. 2, p. 405.

[62] Ibid., p. 89

not readily or vividly recalled." To illustrate the role of competing instincts in the development of conscience, he uses a striking example from the natural world. During the migratory season, female swallows often endure a clash between two very powerful instincts, the migratory and the maternal; a confined bird will beat her breast against the bars of her cage until bloody. The strength and power of the maternal instinct is equally well known, leading even timid birds and animals to risk their lives to protect their young. Yet during late autumn, swallows and house martins frequently abandon their nestlings, leaving them to perish. In this instance the migratory instinct overcomes even the powerful maternal instinct. Now Darwin asks us to imagine the scenario should a swallow gain a human level of intelligence: "When arrived at the end of her long journey, and the migratory instinct ceases to act, what an agony of remorse each bird would feel, if, from being endowed with great mental activity, she could not prevent the image continually passing before her mind of her young ones perishing in the bleak north from cold and hunger." The remembrance of such guilt and pain would undoubtedly impact on the swallow's behavior during the following migratory season. Darwin continues:

> At the moment of action, man will no doubt be apt to follow the stronger impulse; and though this may occasionally prompt him to the noblest deeds, it will far more commonly lead him to gratify his own desires at the expense of other men. But after their gratification, when past and weaker impressions are contrasted with the ever-enduring social instincts, retribution will surely come.[63]

---

[63] Ibid., p. 91.

Darwin makes another, highly significant observation concerning the evolution of conscience that we must take into account before considering early human behavior. At the outset of his description of the evolution of morality, he makes some preliminary remarks:

> It may be well first to premise that I do not wish to maintain that any strictly social animal, if its intellectual faculties were to become as active and as highly developed as in man, would acquire exactly the same moral sense as ours. In the same manner as various animals have some sense of beauty, though they admire widely different objects, so they might have a sense of right and wrong, though led by it to follow widely different lines of conduct. If for instance, to take an extreme case, men were reared under precisely the same conditions as hive-bees, there can hardly be a doubt that our unmarried females would, like the worker-bees, think it a sacred duty to kill their fertile daughters; and no one would think of interfering.[64]

There is a certain contingency to the phenomenon of morality; scientifically speaking, any other social species could have evolved to become as intelligent as Homo sapiens, and therefore another form of morality would have come to predominate on Earth. Chris Stringer, a leading paleoanthropologist, has made a similar point:

> There were plenty of other [evolutionary] paths that could have been taken; many would have led to no

[64] Ibid., p. 73.

> humans at all, others to extinction, and yet others to a different version of "modernity."[65]

Different versions of modernity would have occurred had another hominid species such as the Neanderthals, for example, attained levels of intelligence to equal ours. The notion of evolutionary moral contingency can also, of course, be applied to early Homo sapiens itself. Ruse has made the point that our earliest forebears could have made different behavioral decisions, and hence we could have evolved with different moral sentiments and a resulting substantive morality with a different content to that which actually obtains.[66] This is related to the fact that with the achievement of consciousness and rationality must come the capacity to exercise choice between various behavioral options. This brings me to the deeper and more important question regarding the moral evolution of our species raised at the beginning of the chapter: Does the available evidence bear out the philosophical suggestion that we could have evolved into a morally *superior* species as opposed to a morally different one? Those who agree that the most unattractive qualities of the human species are "hard-wired" by evolution must surely agree with Darwin that our earliest human progenitors, before evolution had begun its selection of behavioral variations, must have been a gentler and less dangerous species than we are ourselves. Darwin, as we have seen, frequently refers to this fact

---

[65] Ibid., p. 273.

[66] Michael Ruse, "Can the Biological Sciences Act As A Ground For Ethics?" in *Contemporary Debates in Philosophy of Biology*, ed. Francisco J. Ayala and Robert Arp (Malden, MA: Wiley-Blackwell, 2010), pp. 297–315. For Ruse, however, any form of human morality would always have had to be regulated by ESSs.

but does not fully explore its implications because he believed that, inevitably, as man "advanced in his intellectual powers [he] retrograded in his instincts."[67] For Darwin, the greatest damage done by natural selection to primeval man was to weaken and sometimes eliminate loving and affectionate familial and social instincts through the continuous use of reason in the pursuit of individual gain. Such individual survival strategies would have frequently involved treachery and violence. Cruel and treacherous behaviors of various kinds thus became part of human nature. The ability to foresee the long-term consequences of one's acts and to feel remorse was not sufficient to prevent the evolution of cruelty, cunning, and the ability to deceive. Throughout *Descent*, Darwin veers between optimism and pessimism concerning the effects of natural selection on human nature. Optimism at the power of reason in enabling man both to understand the consequences of his actions and to feel a duty even toward those who are not members of his immediate family or tribe; pessimism over the loss of loyal and affectionate instincts. Ultimately, pessimism wins out.

The question of whether this evolutionary course was inevitable did not occur to Darwin, probably because he was thinking from a perspective that was radically new and innovative. As I have already pointed out, Darwin was working at a time when the Bible was interpreted literally, and he was already grappling with the huge challenge his theory posed for the Genesis account of creation and the *imago dei*; it is obvious from his writings that the issue of creation was at the forefront of his mind. From philosophical and theistic perspectives however, the question of the inevitability or otherwise of the evolutionary course that has

---

[67] Darwin, *Descent*, Vol. 2, p. 368.

resulted in our particular version of humanity is of the utmost importance. Scientists, philosophers, and theologians who consider the situation of the first humans can easily overlook the fact that we ourselves are the product of a long process of evolution. At the earliest stage of human evolution, when the very first humans had only barely achieved a recognizable humanity and the evolution of the moral sense was just beginning, I would argue that they possessed a power and freedom of the will that is denied to us, their descendants. This greater freedom of the will would have existed in the period of evolutionary instability following the attainment of modern (or nearly modern) intellectual faculties and before the establishment of new evolutionarily stable strategies. The question of whether and to what extent we, their descendants, possess free will is much debated; what is certain is that morally speaking, the will of the first humans would have been freer than ours. Human nature was new, fluid, and untainted. It follows that morality must have been capable of taking various forms, although such forms would of course have been to a large extent constrained and determined by the fundamental nature of our earliest progenitors, inherited from their recent apelike ancestors. Of primary interest to the moral philosopher then are the forms it *might* have taken and why it took the particular form that it did. With this in mind, we will now consider contemporary research in primatology, which has much to add to Darwin's own insights into the behavior of primates that share a common ancestor with the first hominid species and ourselves.

Chimpanzee Cousins

As well as studying anthropological data on primitive hunter-gatherer tribes still or recently in existence, those who try to envisage what early human societies might have been like take into account what primatology has to tell us about the behavior of other ape species, especially chimpanzees. Taxonomically, biologists classify orangutans, gorillas, chimpanzees, and humans as Hominidae, or great apes, and the gibbon as a "lesser ape." Up to the recent past, only humans and protohumans were referred to as hominids, and for the sake of convenience I intend to stick with this earlier means of classification, referring to extinct prehumans as *hominids*, to our remote forebears as *early humans*, and reserving the term *ape* for the four other species of ape.

The current scientific consensus is that orangutans diverged from the human line about fourteen million years ago, the gorilla eight to nine million years ago, and chimpanzees about six million years ago (although this could change); the split between the great apes and gibbons is thought to have occurred fifteen to eighteen million years ago. The basic social structure of gorilla society is the one-male dominated harem, in which the male is polygamous and the females monogamous; this is a closed and stable society and there is no tendency toward larger social groupings, as with chimpanzees. In contrast, orangutans are not social, but solitary and introverted; they live alone and independently as adults, seeking out mating opportunities. The males are loners and range over a large territory, emitting calls as a warning to other males to keep away. Orangutan females are also solitary, except for the powerful bonds they establish with their offspring; they are adapted to raise their children alone and are notably patient mothers. Neither species is territorial, though some orangutans fight for dominance; both gorilla and orangutan males engage in intense competitive aggression over females, however.[68] Gibbons are monogamous, and the only other species of ape besides ourselves to exhibit single pair-bonding; another notable characteristic of gibbons is that the female is the dominant one of the monogamous pair. Of all four species of ape, the structure of chimpanzee society is most like our own; chimpanzees live in social groupings that are very similar to ours, though both sexes are highly promiscuous. On a superficial basis, therefore, it is clear that we have traits of all four species: humans are classified as mildly polygamous, yet mainly pair-bonded. Pair-bonding does not imply the level of monogamy displayed by gibbons, however, since we also share tendencies toward polygamy with gorillas and promiscuity with chimpanzees, though to nothing like the same degree in the latter case. Finally, the solitary preferences and single-parent family structure of orangutans also make up part of the diversity of our species. DNA testing has proven, however, that we share more than 98 percent of our genes with chimpanzees and that they are genetically closer to us than they are even to gorillas.

The divergence of chimpanzees and hominids from their last common ancestor around six million years ago gave rise to many prehuman hominid species that are long extinct; as our closest relatives in the animal world, chimpanzees therefore provide a window on the behavior of now extinct hominids as well as on that of our very earliest ancestors. As a result, they are an important focus of anthropological attempts to determine how extinct hominid species, as well as the earliest members of our own species, would have lived. There are actually two species of

[68] See Paul G. Heltne and Linda A. Marquardt, eds., *Understanding Chimpanzees* (Cambridge, MA: Harvard University Press, 1989).

chimpanzee, each equally closely related to us: Pan troglodyte and Pan paniscus. Interestingly, due to the relatively late discovery of Pan paniscus (more commonly known as the bonobo), it was Pan troglodyte (somewhat confusingly described as the chimpanzee, now often referred to as the common chimpanzee) that from the 1970s onward became the chosen model among primatologists for early hominid behavior. We will consider chimpanzee behavior first, before turning to more recent research on bonobos.

Chimpanzees live in what are known as *fission-fusion* societies made up of permanent, parent communities of about fifty to one hundred; these larger groups split up during the day into smaller, foraging groups that can change frequently in composition. At night, the whole group rejoins and everyone sleeps in the same location. Neighboring communities of chimpanzees are known to fight frequently with one another to expand their territorial ranges, and this leads to a significantly higher death rate among male chimpanzees than among females. Overall, chimpanzee societies are characterized by male dominance hierarchies, social cohesion in hunting and food gathering, and strong family ties. Friendly physical contact is crucial in maintaining good relationships among chimpanzees. For this reason, social grooming (cleaning the hair of another chimp) is their most important social behavior and is a major preoccupation among chimpanzees in a group. Anyone who has observed chimpanzees in the type of modern zoo that attempts to recreate conditions in the wild can easily observe the hours they spend on this pleasurable and relaxing occupation. Chimpanzees also have disputes among themselves that can become physical, and males frequently display aggressive behaviors against the males of other communities. Research that has been carried out on groups of chimpanzees in Africa has found that chimpanzee males who hunt cooperatively for food share the spoils, with the alpha males generally getting more than the others, whether or not they are the most successful hunters. It is also known that alpha males commonly use food acquired in the hunt for political purposes to maintain their positions of dominance and to reward sexually available females. Many chimpanzee hunting parties display complex hunting techniques and tactics that demand a high level of coordinated activity; these techniques can take several years to learn. Those who have been unsuccessful in the hunt but hold out their hands in a "begging" gesture will usually be given meat. Researchers have also observed other interesting behaviors related to hunting for food and danger from predators. For example, chimpanzees communicate with a wide range of postures and gestures. Their food calls—grunts, barks and pant hoots—alert other chimpanzees to the whereabouts of food; a special intensity of excited calls indicates that there has been a successful kill or that an abundance of food has been found. Chimpanzees invite one another to feast. As Darwin pointed out in his discussion on the evolution of language, chimpanzees, when they come across something unusual or dangerous, emit danger cries to warn other chimps in the area; their danger call can be heard through the forest for about two miles. Chimpanzees also display human-like violence in strategically cooperative male battles between rival groups over females and territory; further, males are notably dominant over females, and sometimes subject them to physical brutality. Regarding intelligence and reasoning powers, research proves that Darwin was correct in attributing a high level of both to chimpanzees: studies carried out in the wild by contemporary primatologists such as Richard Wrangham, Jane Goodall, and Toshisada Nishada have shown, for example, that chimpanzees often use tools; they have been observed using sticks to obtain ants and termites to eat and to scare away intruders. They also use chewed leaves like a sponge to sop up water to drink. It is a well-documented fact that chimpanzees in East Africa treat themselves when they are sick by using the same plants—*Aspilia*—that local people use for the identical illnesses. Groups of chimpanzee males, as we have seen, display complex and coordinated hunting strategies.[69]

Chimpanzees are so human-like in many of their characteristics—particularly in cooperative male warfare and dominance over females, it must be said—that primatologists were taken by surprise in the 1980s when research on bonobos in captivity and in the wild began to bear fruit; the differences between bonobo and chimpanzee societies have broadened out our perspective on hominids and early humans in a way that challenges all presumptions and preconceptions about how early human societies might have operated. In comparison with chimpanzees, bonobo society is relatively egalitarian; although there are dominance hierarchies, they are not so rigid, and the dominance and status that does exist apparently relates more to female coalitions than to male ones.

The main differences between the two species of ape began with their physical characteristics. Leading primatologist Frans de Waal has commented:

> In physique, a bonobo is as different from a chimpanzee as a Concorde is from a Boeing 747. . . . The bonobo, with its long legs and small head atop narrow

---

[69] Ibid. See also the Jane Goodall Institute's website (www.janegoodall.org) for regular updates on research.

> shoulders, has a more gracile build than does a chimpanzee. Bonobo lips are reddish in a black face, the ears small and the nostrils almost as wide as a gorilla's. These primates also have a flatter, more open face with a higher forehead than a chimpanzee's and—to top it all off—an attractive coiffure with long, fine, black hair neatly parted in the middle.[70]

Their graceful physiques, small heads and relatively long legs mean that the elegant bonobos are physically closer to humans than are the stockier chimpanzees.

De Waal's interest in bonobos came about because of his research on aggressive behavior in primates. In relation to chimpanzees, bonobos are remarkably peaceable and tolerant; there appears to be little intracommunity or intercommunity conflict. De Waal made the fascinating discovery that the risks of violence over food and the acquisition of mates are greatly lessened in bonobo society due to the use of sex as a means of diffusing tension in competitive situations. Situations that lead to fights and retaliation in most other primate species (including our own) are averted by the use of sex to both avoid conflict and effect reconciliation when it does occur. For bonobos, sexual and social behaviors are indistinguishable, and this is thought to explain their peacefulness and tolerance. Bonobos as a species are also highly bisexual, with frequent sexual interactions occurring among the females in particular. Further, bonobo females are more like human females than chimpanzee females

---

[70] Frans B. M. de Waal, "Bonobo Sex and Society," *Scientific American*, 272(3):82–88. See also Heine and Marquardt, *Understanding Chimpanzees*, pp. 152–207.

in being almost permanently sexually receptive; chimpanzee females—as is the general rule among mammalian females—are only receptive during estrus, which is associated with the time of ovulation. A related psychological similarity between humans and bonobos is their ability to engage in recreational as opposed to mainly reproductive sex. Finally, bonobos are frequently observed kissing one another in a strikingly human-like way.

The relative—in chimpanzee terms—egalitarianism of bonobo societies is largely due to the fact that the females in a troop form powerful alliances that seem to give them overall control of the group. It is known, for example, that bonobo males remain attached to their mothers throughout their lives and depend on them for protection in their encounters with other males. Interestingly, in bonobo communities the highest-ranking males are usually sons of the most important females. This is in striking contrast to chimpanzees, which either fight their own battles or rely on the support of other males. Even more remarkably, bonobo society seems to be not only female bonded, but female dominated. Research on feeding practices has shown that when an unexpected portion of food is given to a group of chimps composed of two females and one male, the male chimpanzee will make a charging display through the enclosure and claim the food for himself; only when he is satisfied are females allowed near the leftovers. In identical bonobo groups, the reverse happens: the females approach the food first, and even if the male makes repeated charging displays, he is ignored. In other experiments, it has been reported that if a male bonobo tries to harass a female, all the females in the group band together and chase him away. Research indicates that females are better able to dominate males when they are in a group and that their close bonds are most likely a means to forming alliances; in contrast, little female bonding has been observed among chimpanzees. The relative egalitarianism of such female coalitions also sharply contrasts with the hierarchical, competitive, and socially distinct male coalitions formed by chimpanzees. Whereas the males of different chimpanzee communities generally fight with one another—with lethal consequences for many—peaceable mingling that includes sex and grooming has been observed between different bonobo communities; serious conflict does occur, but is apparently rare. The latter state of affairs is thought to be due to the domination of young bonobo males by coalitions of older females. Such female solidarity over against the bonobo males is all the more remarkable since bonobo females—like chimpanzee females—must leave their home troop and be accepted into a different one in order to mate and produce offspring. Male bonobos and chimpanzees, by contrast, are brought up together and remain in their home troops throughout their lives; before research was carried out on bonobos, it was felt that chimpanzee male coalitions could be partly explained by the fact that the males had known one another since birth, while the adolescent and adult females came from different troops and had to start off as strangers. Yet female bonobos are able to quickly establish enduring bonds with one another. Finally bonobos, like chimpanzees, are highly intelligent though far more sensitive in temperament. De Waal reports that during the World War II bombing of Hellabrun, Germany, all of the bonobos in a nearby zoo died of fright from the noise while the chimpanzees were unaffected.[71] Overall, however, chimpanzees and bonobos share far more similarities than differences; they are characterized by attractive,

[71] Ibid., p. 3.

outgoing personalities and, as we have seen, share similar fission-fusion social systems. Of some significance is the fact that the greater territoriality and aggression shown by chimpanzees is extremely mild when compared to that of humans. What are the implications of the behavior of our closest relatives in the animal world for our behavior? In the following chapter, we will build on Darwin's insights as well as the above research in primatology by turning to the evidence of paleoanthropology and archaeology to trace the evolution of prehuman hominid species into early modern humans.

Chapter 4
The First Humans: A Gentler Species

New methods of dating and interpreting fossil finds in the field of paleoanthropology are currently revolutionizing our knowledge of the social behavior of prehuman hominids. Even without this knowledge, however, it would be safe to argue that the strong similarities between bonobo, chimpanzee, and human behaviors, allied to the fact that modern humans share traits of both ape species that the latter don't share with one another, indicate that the earliest hominids are likely to have been a combination of both, with added characteristics that are unique to the genus *Homo*; it is the latter line with which we are mainly concerned in this chapter. We now know that the very earliest representatives of our own species lived at the end of a period called the Miocene and at the beginning of the following Pliocene era, between 6 and 4.5 million years ago. The Miocene—a kind of "golden age of the apes"—lasted from about 22 million years to about 6 million years ago; scientists have identified more than twenty types of extinct apes from sites in Africa. The earliest of these ancient apes are known as *proconsuloids* and were more similar to monkeys than to their present-day descendants the great apes, as described in the previous chapter. Around 16 million years ago, the proconsuloids had evolved into the "hominoid" apes that would split into the line leading to chimpanzees, bonobos, gorillas, and orangutans on the one hand, and a diverse group of our own hominid predecessors on the other.[72] Of most relevance to our discussion is the behavior of the first hominid types and especially how such behavior differed from that of their closest relatives, the hominoid apes. Because behavior in the natural world is directly related to biology, we will first consider the differences in physique between both species and what the implications of those differences might be for the evolution of hominid behavior.

The *Australopiths*: Pre-Hominids

The discovery of the fossil remains of the precursors of our species is so recent that there is still much confusion and disagreement among paleoanthropologists as to how they should be classified. This is particularly so in relation to the *Australopiths*, the very earliest species to be firmly identified as our ancestors; they are variously classified as bipedal apes or the first members of the genus *Homo*. *Australopiths* appeared around four million years ago, and combined ape and hominid features; their upper bodies, skull proportions, and faces were apelike, while their lower bodies were adapted to upright walking. The latter ability—to walk upright—allowed them to leave the protection of the forest, where chimpanzees live in the trees, for the forest-edge and woodland areas that bordered the savanna; this meant that they could exploit the food resources of both forest and savanna. The new advantages and possibilities came at a price, however: the open habitat of the grasslands would have rendered these

---

[72] Ian Tattersall, *Masters of the Planet* (New York: Palgrave Macmillan, 2012), pp. 2–3.

small-bodied bipeds more vulnerable to predators than when they had lived exclusively in their ancestral habitat in the deep forest. Paleoanthropologists believe that the process of adapting to this new situation would have led to the evolution of significant new behaviors and that these behaviors would have enhanced the already social nature of the *Australopiths*. Ian Tattersall, for example, turns to the lifestyles of more distantly related though environmentally similar modern primates such as baboons and macaques for indications of the likely behavioral changes that would have occurred. Baboons and macaques also exploit both forest and savanna, seeking protection from predators by living in very large groups of males and females of various ages, all alert to the danger and ready to raise the alarm when necessary. They are also, unsurprisingly, highly vocal. Tattersall speculates that the *Australopiths* would have lived in larger social units than the average-size chimpanzee ones for protection, and that during this period of their evolution they would have stayed together all the time rather than splitting into smaller groups during the day, as in the forest environment. Because of the need for mutual protection, they would, he speculates, have had complex social lives rooted in intense forms of cooperation and would also have been highly vocal due to the need for protection from predators. Tattersall argues that "by analogy with living apes . . . they would have had a vocabulary of several dozen distinct utterances, each expressing one of a range of different situations or emotional states."[73]

Apart from such likely social advances, it is known beyond doubt that the *Australopiths* had made a cognitive leap that places them far beyond chimpanzees in intelligence. As early as

---

[73] Ibid., p. 61.

2.6 million years ago, they were using tools (classified as Oldowan after Olduvai Gorge in Tanzania, where such tools were first found) to butcher the carcasses of animals; these were small cobbles of carefully chosen fine-grained rock that were hit with another stone to detach sharp flakes for cutting. The evidence is also clear that these ancient bipedal apes carried appropriate rocks around with them in anticipation of successful scavenging or hunting, and they understood which types of stone would be best for their requirements. Such behavior indicates a level of craftsmanship, foresight, and planning that is far beyond the capabilities of even the most highly trained modern chimpanzee, as controlled experiments have shown. There is also some evidence to suggest that *Australopiths* had learned how to throw heavy objects at larger prey. Throwing—as Tattersall points out—seems perfectly normal and natural to us today, but it is one of the quintessential skills that distinguishes us from all other species, including our closest relatives, the chimpanzees.

One of the behaviors that marks a highly significant difference between chimpanzee and hominid species began to evolve during the era of the bipedal apes. As a species we are classified as mildly polygamous; monogamous pair-bonds are not only commonplace but held up as the ideal in many societies. As such, the amount of time and effort the human male is prepared to expend on his offspring is unique among primate species. Thus, a major difference between early humans and their apelike predecessors would have been the evolution of male involvement in the rearing of children and a consequent lessening of sexual promiscuity. While there are strong similarities between chimpanzee and human sexual behaviors, the chimpanzee "family" (whether the common chimpanzee or the bonobo) consists of mother and children, with the females in a troop living together and the males having no input into child-rearing. In addition, female chimpanzees and bonobos are highly promiscuous relative to human females in that they mate indiscriminately during estrus. How, therefore, might such an evolutionary leap have occurred? Despite their promiscuity, some aspects of bonobo sexuality, as we have seen, show similarities to human sexual behavior that are unknown in other primate species. Prolonged female sexual receptiveness may have been a characteristic of the *Homo* line after the separation of chimps and hominids, facilitating the evolution of monogamy; the latter aspect of bonobo sexual behavior is significant enough to raise the possibility that the last common ancestor of chimpanzees and humans may have been closer to bonobos than to the common chimpanzee. This possibility is strengthened by the fact that the gracile *Australopiths*, considered to be the ancestors of all hominid species, are biologically more similar to bonobos than to chimpanzees. If however the *Australopiths* started out with sex lives similar to those of bonobos, dramatic changes would have had to occur in order for the human nuclear family to evolve.

Paleoanthropologists believe that pair-bonding began in earnest after the separation of the ape and hominid lines and that the diminution of polygamous and promiscuous instincts was a first, major step in the evolution of human behavior. It is thought that intensive male involvement in the rearing of children originated as a result of biological changes that occurred when early hominid species were making the transition from tree climbing to bipedalism. As apelike lower limbs evolved to support and propel an upright body, the opposable thumb-like big-toe characteristic of ape species would have been lost, with momentous consequences. Chimp babies are able to cling to their mothers' hair with muscular long-fingered hands and grasping toes; this leaves the mothers free to use their hands for gathering food while protecting themselves and their infants from danger. Among the *Australopiths*, the opposable big toe was gradually lost with the result that infants would have had to be carried by their mothers, thereby limiting the mothers' ability to provide for and protect themselves and their young. This may have meant greater dependence on a single mate and acted as a spur to the evolution of pair-bonding.

Another significant factor was the evolution of the long hominid childhood; chimpanzee young begin to forage for themselves as soon as nursing has ended, at about four years of age. The length of hominid childhood is unique among mammals; it is a direct result of the large hominid brain, the time it takes to develop, and the consequent vulnerability of hominid infants. Fossil remains show a gradual increase in the length of hominid childhood that occurred in tandem with increasing brain size, beginning with the three-million-year-old *Australopithecus afarensis* and continuing to our own species in which childhood evolved to be slightly longer than in the Neanderthals.[74] Over time, this would have led in

---

[74] This can be deduced from the time of eruption of molars estimated from the fossil skulls of hominid species. It is known, for example, that the Neanderthal childhood was slightly longer than that of *Homo erectus*, though shorter than that of modern humans. See Stringer, *The Origin of Our Species*, pp. 62–64, 72. See also Christopher P. Sloan, "The Origin of Childhood," *National Geographic*, November 2006, which documents the discovery in Ethiopia's Rift Valley of the fossil remains of a baby *A.afarensis* whose milk teeth and unerupted adult teeth were perfectly preserved. The baby's remains were found in 2000 in a region named Dikika, a hill area just across the Awash River from Hadar, the site in Ethiopia's Rift Valley, where the most famous *A. afarensis* fossil, a 3.2-million-year-old female known as Lucy, was found in 1974.

hominid species to a longer period of involvement from the male parent and the gradual establishment of monogamous pair-bonds facilitated by the evolution of continuous female sexual receptivity. Early prehuman hominids would therefore have been far less promiscuous than their recent apelike ancestors and also less promiscuous than us, their fully evolved descendants since, as Darwin pointed out, the evolution of the nuclear family would have been reinforced by habitual instincts. They would, in all likelihood, have been even more peaceful than bonobos. Because hominid males would have had permanent mates, there would have been no need for the type of competition over females displayed by gorillas, orangutans, and chimpanzees; nor would there have been any need for the constant, promiscuous sexual interactions that bonobos engage in to reduce tensions and avoid conflict. Hominid males would have had noncompetitive, regular sexual interactions through monogamous mating and pair-bonding, with an accompanying reduction in aggressive competition over females and resources.

The transition from *Australopith* to fully fledged *Homo* was marked by full bipedalism and life in the open. At least ten early species of *Homo* have so far been identified, dating from about two million years ago; their classification remains fluid in the context of ongoing archaeological and paleoanthropological discoveries. It is widely accepted, however, that one of them, described as *Homo erectus*, is the earliest direct ancestor of *Homo sapiens* so far found. *Homo erectus* evolved in Africa almost two million years ago and marked a benchmark change in hominid evolution, possessing a brain size beyond any ape or australopithecine, a human[-like] face with projecting nasal bones, small teeth, a human-like posture for the skull, and a body frame of human rather than ape-like proportions.[75]

Homo erectus was probably the first species of hominid to engage in systematic hunter-gathering and to gain control of fire; fossil remains reveal that as the emerging hominid species moved further away from their apelike ancestors, their teeth grew gradually smaller, indicating that they had learned how to process and make food digestible through cooking it. By 1.5 million years ago, *Homo erectus* had made a significant technological advance by learning to make and use teardrop-shaped hand axes, which are usually classified by the term *Acheulean*. These were a significant advance on Oldowan technology, being much larger at eight or nine inches long and far more elaborate. A much higher degree of dexterity and skill was required to shape the symmetrical, teardrop form; moreover, as well as having to choose the right kind of rock (as the Oldowan toolmakers did), the hand-ax makers also had to ensure that the individual piece of rock chosen was free of flaws that would hinder the flaking sequence needed to shape it as required. Hence the toolmaker would have had to be able to envisage the "form" of the tool in the piece of rock and to ascertain that the core could support a complex sequence of manipulations. The hand axes were used for a variety of tasks including meat cutting and the scraping of hides; they were made in large quantities in "workshop" locations, which strongly suggests role specialization within the group. Such an innovation must have had profound implications for the evolving cognitive abilities of *Homo erectus* and other, similar species.[76]

---

[75] Chris Stringer, *The Origin of Our Species* (London: Allen Lane, 2011), p. 28.
[76] Ian Tattersall, *Masters of the Planet*, pp. 125–126.

Erectus was also the first hominid species to travel out of Africa, and it is in relation to some of its fossils, found in the abandoned medieval town of Dmanisi in the Republic of Georgia, that the first indications of hominid empathy have been found. Dmanisi was being excavated by archaeologists seeking to uncover new aspects of life along a branch of the medieval Silk Road trading route when, unexpectedly, the fossil tooth of a rhinoceros turned up. As it happened, the town had been built on a much more ancient fossiliferous basalt deposit; the fossils found included the remains of five small-brained hominids dating back 1.75 million years. Although this little band of hominids was still using Oldowan technology, its members seem to have developed empathy and compassion beyond anything possible even for empathetic chimpanzees. One of the five fossilized skulls belonged to an elderly individual and contained just one tooth. All of the empty tooth sockets had shriveled away, a lengthy process during which the old male would have starved to death without a great deal of routine assistance from his social group; this seriously disadvantaged individual was evidently loved and supported by his relatives. While chimpanzees have been seen to comfort injured or oppressed members of their group, what they lack in comparison to the Dmanisi hominids is the technical ability to give practical assistance.[77]

That first wave of *erectus* is thought to have died out or evolved into other forms. About 1.5 million years ago, African *erectus* moved out of Africa a second time into Asia and Europe, where, about 500,000 years ago, it possibly evolved into a species known as *Homo heidelbergensis*. A similar evolution occurred in Africa around the same time, and it is likely that the resulting

---

[77] See Stringer, *Origin*, pp. 77–78; Tattersall, *Masters*, pp. 119–123.

line—also *heidelbergensis*—had evolved into the ancestors of modern humans sometime between 300,000 and 130,000 years ago. *Heidelbergensis* had a brain size around double that of *erectus*, within modern human range but below the average. These hominids had made further social and cognitive advances on those of their predecessors, as evidenced at numerous sites. A particularly interesting refuge and resting place has been found on what then would have been a beach at Terra Amata, a suburb of Nice. Here, at about 380,000 years ago, a group of hominid hunters had built shelters made of saplings to which, it seemed, they regularly returned after hunting expeditions. The best preserved of these shelters had marked out entrances with hearths built directly inside, so that the smoke could escape; they contained the bones of animals that had been caught and cooked. Although evidence of the control of fire had already been found at an 800,000-year-old *erectus* site in Israel, the Terra Amata site signals the use of fire for cooking and warmth as a regular part of documented hominid behavior. The role of fire in the evolution of hominid—and hence human behavior—cannot be overestimated. As Darwin himself pointed out, "The art of making fire . . . is probably the greatest [discovery], excepting language, ever made by man."

> It provided warmth and protection from predators, illumination to extend "daylight" and a new social focus as people sat to talk, sleep or work (and later to sing and dance) around the flickering flames.[78]

[78] See Tattersall, *Masters*, p. 138, and Stringer, *Origin*, pp. 139–141.

Anthropologists believe that female bonding and cooperation in the collecting, gathering, and processing of foods for cooking was critical to the catalyzing of social change in early prehuman hominids. Moreover, the pleasures of sitting around the fire for food and warmth would have encouraged bonding, sociality, and even—at the risk of anthropomorphizing—a primitive form of conviviality. Most of all, perhaps, the work of women in preparing and cooking food over open fires, followed by the sharing of food, would have provided further impetus for the development of some form of language in even the earliest hominid tribes.

As might be expected from the doubling of its brain size, *Homo heidelbergensis* had also made significant advances in hunting and toolmaking. At Schöningen, in Northern Germany, several long, javelin-like, beautifully crafted wooden spears made of yew and spruce—around 400,000 years old—were found along with the skeletons of about twenty horses. The use of such sophisticated spears in the hunting of large wild animals by relatively puny hominids highlights the level of courage, cooperation, and cunning that had come to characterize hominid behavior at around a half million years ago.[79]

Neanderthal Brothers

Between 300,000 and 400,000 years ago,[80] *Heidelbergensis* had begun to evolve into Neanderthals in both Asia and Europe. The Neanderthals (*Homo neanderthalensis*) are the species about which most is known. They are of particular interest for three reasons: they were the last hominid species to become

[79] Stringer, *Origin*, p. 109.
[80] Ibid., pp. 25–26.

extinct, they have left behind a rich fossil record, and they coexisted with modern *Homo sapiens* in Europe for at least ten thousand years.

Neanderthals evolved outside Africa and appear to have roamed Europe and southwest Asia unchallenged from about 200,000 to about 40,000 years ago. The fossil record shows that they had developed a tool technology—known as Mousterian—far more sophisticated than that of the Acheulean; their repertoire included about sixty different types of flake tool and a large quantity of different hand ax types. A major innovation was the ability to shape stone templates, or "prepared cores," from which tools of various types could be struck. According to Tattersall, the Neanderthals were gifted stoneworkers who sought out and highly valued good materials:

> The Neanderthals instinctively knew stone, as a modern cabinetmaker instinctively knows wood . . . The [Mousterian craftsmen] carefully fashioned a good piece of flint or chert, then gave it a new edge over and over again until it was too small to be of further use.[81]

It is also known that the Neanderthals fashioned composite tools by setting their stone implements into wooden handles; tellingly, however, they rarely made tools of other soft materials such as bone and antler, though the latter have been found in plentiful supply at Neanderthal sites. Bone and antler were habitually sculpted into tools and artefacts by the later, fully modern *Homo sapiens*, indicating a qualitative gulf between

[81] Tattersall, *Masters*, p. 176.

Neanderthal intelligence and that of *Homo sapiens*. The fact that their brains were as large—and sometimes even larger—than ours does not seem to imply similar cognitive or symbolic abilities.

The Neanderthals lived in small groups that—as we will shortly see—were probably based on kin relationships. They were highly carnivorous, more robustly built than ourselves, and had stronger bones; they were also courageous and proficient hunters who engaged in cooperative, confrontational hunting at close range with fearsome large-bodied animals such as mammoths and wooly rhinoceroses. Like their predecessors since time immemorial, they led a wandering existence, moving from one place to the next as they used up local resources. The Neanderthals had control of fire and were the first hominid species to bury their dead; it has even been suggested that objects such as special rocks or tools found at burial sites may have been grave goods. Although they left behind little that would suggest belief in an afterlife or any capacity for symbolic thought, the fact that they buried their dead with objects beloved by their owners in life certainly indicates that they must have had deep feelings for them. There is, in fact, an abundance of emerging evidence for empathy and caring among Neanderthals. Their fossil remains show that they sustained many injuries—probably from hunting—that they could not have survived without social support from their tribes. At the Sima de los Huesos site at Atapuerca, Spain, dating from about 400,000 years ago, the remains of a child with a deformed skull and brain who would have been mentally and physically disabled was obviously not rejected at birth since it died naturally at around the age of eight. Further evidence of Neanderthal compassion was found at Shanidar, a site in northern Iraq; the skeleton of a Neanderthal male excavated there had healed injuries to his head and showed a withering of the right arm that might have been with him since birth. The Neanderthals had to survive in environments that were frequently harsh due to ice-age climatic conditions during much of their existence as a species; in such a world, an injured or disabled individual could not have survived without active care and support from his social group.[82] Most interesting with regard to the Neanderthals is evidence that they lived in small familial groups, in contrast to the societies of chimpanzees, in which the females take care of offspring and the males consort primarily in coalition with one another. Up until quite recently, there was little evidence to show how Neanderthal groups were organized, although the evidence that does exist indicates that the groupings were small, ranging from eight to ten individuals; the recent discovery of the fifty-thousand-year-old Spanish site of El Sidrón has, however, provided us with a larger window into the structure of Neanderthal society.

The El Sidrón site contained the fossil remains of a Neanderthal grouping made up of twelve Neanderthal individuals: six adults, three adolescents, two juveniles, and an infant. The genetic evidence indicates that this was an extended family. The three adult males were brothers, while the females were unrelated to one another but were related to the three children; it is thought that the adolescents were probably related in some way to the adult males. There are, however, similarities to both chimpanzee society and those of modern hunter-gatherers in this scenario. As we saw from the previous chapter on chimpanzee and bonobo behavior, the females leave their own band

[82] Ian Tattersall and Jeffrey Schwartz, *Extinct Humans* (New York: Westview Press, 2000), p. 216. See also Stringer, *Origin*, pp. 148–149.

and join others to mate and produce offspring while the males remain in their natal groups. This social system is known as *patrilocality* and is important in reducing inbreeding; it is also a common one among modern hunter-gatherers—and indeed much later on in monarchical systems—and hence another form of behavior that is shared by Neanderthals and moderns. The familial implications of such small groupings are further strengthened by the fact that Neanderthal males and females show low levels of sexual dimorphism; this means that, like gibbons, they may have been pair-bonded and fully monogamous. It is thought that the greater the bodily dimorphism (differences in size) between the sexes in a species, the greater the likelihood of intense male competition over females and of increased rates of dominance and polygamy among males. Gorilla males, for example, are far larger than monogamous gorilla females. There is further evidence for monogamy among Neanderthals in the fact that Neanderthal skeletons, whether of men, women, or children, display a high frequency of lesions and fractures that are consistent with the hunting of big game; their skeletons would seem to indicate that hunting was a family affair, with wives and children perhaps acting as drivers or beaters.[83] While the evidence is far from conclusive, it is possible that Neanderthals were an even more family-oriented species than *Homo sapiens*, with less male dominance and role specialization in relation to sex. *Homo sapiens* displays a greater degree of sexual dimorphism than *Homo neanderthalensis*, reflecting a higher degree of male dominance, polygamous tendencies, and also, it must be acknowledged, potential for higher levels of attraction between the sexes.

---

[83] Ibid., p. 143. See also Tattersall, *Masters*, pp. 171–175.

The El Sidrón fossils also, however, indicate a darker side to Neanderthal existence; for it seems—although the evidence is compelling rather than conclusive—that the members of this particular little grouping had been massacred and cannibalized; many of the bones show cut marks consistent with defleshing. There are various theories as to why such behavior might have occurred; defleshing may, for example, have been part of some ancient funeral rituals and not necessarily proof of cannibalism. If cannibalism did occur, it might have been for the purposes of survival during a time of catastrophe or starvation, something that has also occurred in recent human history. Interestingly, the El Sidrón fossil remains bear signs of dietary stress, indicating malnutrition. Evidence of cannibalism stretches back to the earliest hominid species however, including the one thought to be the immediate precursor of our own; we also know, of course, that the butchery and consumption of human flesh has occurred throughout the timespan of *Homo sapiens* on Earth, up to the recent past.

Were Neanderthals and similar hominid species able to speak? We have not yet addressed the issue of hominid speech beyond speculation on vocal development among *Australopiths*; we will return to the subject shortly in more detail. The Neanderthals are particularly interesting in this regard since they had brain sizes as large as our own. While their vocal tracts were closer to those of a two-year-old child than a modern adult's, this would have been sufficient for at least simple vocal communication. Stringer believes that simple forms of language must have existed in even earlier hominid species than the Neanderthals, given the complexity of behaviors and technology that is apparent at Early and Middle Stone Age sites such as the aforementioned Schöningen (400,000 B.C.) in Germany, Boxgrove

(500,000 B.C.) in England, and Kapthurin (250,000 B.C.) in Kenya. It seems logical that cooperative behavior in hunting and tool-making would have required some form of language. The Mousterian tool technology of the Neanderthals represented a major advance in stoneworking that marked the transition from the Early to the Middle Stone Age period (200,000–45,000 B.C.), and there is no doubt that they were highly knowledgeable about the world in which they lived; so it is more than probable that they could communicate vocally with one another.[84] Stringer speculates, however, that this would have been a language of the "here and now," without future or past tenses, symbolism, or abstract reasoning.

Not Yet Human: Anatomically Modern *Homo Sapiens*

Although the archaeological evidence reveals that prehuman hominids had achieved far higher levels of intelligence than had their chimpanzee cousins, it also thus far reveals an incalculable gulf between hominid and human intelligence. From around forty thousand years ago, at about the time modern *Homo sapiens* arrived in Europe, archaeological evidence from sites in France and Germany shows an explosion in the development of intelligence. By the time they reached Europe, our ancestors (known as the Cro-Magnons) had developed the ability to

---

[84] Boxgrove is a British archaeological site near Chichester that has produced more than 400 beautifully made flint hand axes and evidence of advanced hunting and butchering techniques; this evidence has led to "a re-evaluation of the capabilities of hunter-gatherers 500,000 years ago." The Kapthurin site in Kenya shows that the use of red pigment goes back further than 250,000 years in Africa, which raises the possibility that those who used it had symbolic awareness. See Stringer, *Origin*, pp. 66–69, 210, 109, and 158–159.

engage in sculpture, engraving, painting, music, elaborate burials, and specialized social roles; they were us, minus the gloss that civilization can bestow. This advance in intellectual powers appears to have been extremely rapid by evolutionary standards; the selective pressures that drove it are not yet known, although most experts believe that those pressures originated in Africa somewhat earlier. At any rate, the evidence indicates that our species emerged in Africa between two hundred thousand and three hundred thousand years ago. Unfortunately, the fossil record for this period is—at least at this point in time—nowhere near as rich as the European one, which means that a great deal of paleoanthropological and archaeological detective work has been required to piece together some semblance of a narrative that can convey the lifestyle and evolutionary trajectory of the earliest members of our species.

The very first traces of individuals whose bony structure would have matched our own were discovered in two locations in Ethiopia. A skull—known as the Omo skull after southern Ethiopia's Omo basin, the location in which it was found—is thought to be around one hundred ninety-five thousand years old. Three crania were subsequently found in Herto, in Northern Ethiopia, dated to between one hundred fifty-five and one hundred sixty thousand years old. The features that mark the Omo and Herto skulls as unequivocally early members of our species are those that distinguish us clearly from all other hominid species who preceded or were contemporaneous with us. The cranial anatomy of *Homo sapiens* is unique in its high, globular, and voluminous vault, underneath which is tucked a relatively small face. Other hominid species, such as the Neanderthals, had large faces that receded backward into sloping skulls. Stone tools found at the sites, however, ranged from Acheulean to the equivalent of Mousterian, indicating that these anatomically modern looking hominids had not attained the skills of their European descendants some one hundred twenty thousand years later. Other fossil hominid crania that resemble ours have been found in Africa following the Omo/Herto time period, also associated with Middle Stone Age technology. However, the most spectacular and interesting fossils have been unearthed in the Israeli caves of Jebel Qafzeh and Skhul. Two more or less complete skeletons—one adult, one adolescent—dating to one hundred thousand years ago and clearly members of our species, were found in the Qafzeh cave, along with evidence of more than a dozen *Homo sapiens* burials. In the nearby Skhul Cave, an early modern *Homo sapiens* was found to have been interred holding the jaw of an enormous wild boar at around one hundred fifteen thousand years ago. Some experts claim that the latter burial is the first so far found that displays symbolic behavior, which would imply an intellectual gulf between early modern humans and other hominid species such as the Neanderthals. This is a debatable point, however, since in other respects the early moderns of the Skhul and Qafzah caves behaved like other Middle Stone Age hominids. Paleoanthropologists tend to distinguish between what they call "anatomically modern humans" and behaviorally modern ones, given that humans who were anatomically similar to ourselves seem to have shared the same technology as Neanderthals and other archaic hominid species for thousands of years. What makes these fossil finds particularly interesting is that Neanderthal fossils have also been found in the same caves, indicating that the two different species of hominid coexisted. The geographical area in which the caves are situated is known as the Levant, an area along the coast of the Mediterranean Sea and intermediate between Africa and

Europe; it would have been the first stop for hominids leaving Africa, but would also have provided refuge for Neanderthals escaping Ice Age conditions in Europe. Hence there appears to have been an ebb and flow of anatomically modern *Homo sapiens* and Neanderthal populations sharing the same food resources, cave shelters, and technology for a protracted period that lasted from about one hundred thousand to about sixty thousand years ago.[85]

Other than possible symbolic capacities, was there anything else about these early modern humans to distinguish them in such a way from other Middle Stone Age African hominids as to provide a stronger intimation of what was to come? Much has been made of the unique cranial shape of *Homo sapiens*, along with his large brain. It must be kept in mind, however, that anatomically modern humans with large brains shared the same Mousterian-type tool technology as the Neanderthals for the best part of two hundred thousand years, and that the Neanderthals had brains as large as, or larger than our own, without ever developing our level of cognitive ability. The relevant authorities speculate that the crucial difference must relate to the inner wiring of the brain rather than brain size. There is, as yet, no satisfactory evolutionary explanation—either graduated (Darwinian) or punctuated (Lamarckian)—for the appearance of the full suite of modern human behaviors displayed by the Cro-Magnons. It is of course possible that anatomically and behaviorally modern traits emerged simultaneously and that the cognitional gulf occurred between the first humans of our species and an earlier archaic one from which they evolved, rather

[85] Tattersall, *Masters*, p. 186; Tattersall and Schwartz, *Extinct Humans*, pp. 224–247. See also Stringer, *Origin*, pp. 42–44.

than within the species itself. The apparently sudden appearance of the full suite of human cognitional attributes some forty thousand years ago may have been due to a prior rapid cultural evolution. For the cultural transmission of new skills and discoveries to occur, human populations would have had to achieve high densities, and population increases may have been the crucial factor in the rapid cultural evolution of Cro-Magnons rather than a qualitative leap in intelligence. The crucial cognitive advancement may have occurred—and indeed been related to—the attainment of our modern anatomical form. The latter remains a hotly debated question.

A major feature that distinguishes us from other hominid species is, of course, language, a highly symbolic activity. While experts in the field such as Stringer argue that Neanderthals and their equivalents must logically have had some linguistic capability, they would not have the symbolic capacity for language as we know it. Neanderthals did have a gene called *FOXP2* that is known to be intimately involved in language abilities; it transpires, however, that there are numerous genes involved in the developmental process that leads to human speech, quite apart from the neural ingredients. The Neanderthals possessed a necessary condition for language and certainly for rudimentary language, but that is most likely as far as it went. It is in this regard that one aspect of the morphology of anatomically modern *Homo sapiens* assumes a certain significance; alone among other documented species, we possess a low larynx, which makes it easier for the throat muscles to produce the sounds necessary for sophisticated speech. Tattersall speculates that however and in what manner our species emerged in its fully modern form, its first members must have "belonged to an ancestral society that already possessed an elaborate system of interindividual communication, one that employed vocalization as well as gesture and body language."[86] An interesting point in relation to our vocal equipment is the human love of music and singing. Did hominids sing? If so, were anatomically modern humans the most proficient hominid singers? We will discuss this question in more detail in the following chapter.

The abilities necessary for complex language correlate, of course, with the ability to fashion complex tools and symbolic artefacts, objects found in abundance in Upper Paleolithic Europe from about forty thousand years ago. While nothing has been found thus far in the Early to Middle Stone Age African record to match the achievement of Late Stone Age Africans or the European moderns of the Upper Paleolithic (the equivalent of the Late Stone Age in Africa), there are some indications at certain African Middle Stone Age sites of a degree of symbolic behavior. At Skhul, for example, two perforated *Nassarius* tick shells were found, dating from one hundred thousand years ago, that may have been used for personal decoration—a symbolic activity. Although the evidence is tenuous, it is considered to be of some significance that the same shells were used as beads in later, more substantiated symbolic contexts. At about seventy-six thousand years ago at the South African coastal Blombos Cave, archaeologists found an abundance of perforated *Nassarius* shells that are thought to have been strung as beads and used for personal adornment. There is even more explicit evidence for symbolic thought from Blombos in three ochre plaques crafted during the same timeframe as the beads and engraved with a distinctive pattern. Blombos Cave has also yielded tools that were made using a pressure-flaking technique some

[86] Tattersall, *Masters*, p. 220.

fifty-five thousand years before it turned up in the Cro-Magnon *Solutrean* industry of Europe. Significantly, bone tools were also found at Blombos, indicating that the hominids who hafted them were superior in terms of craftsmanship and cognition to Neanderthals. At another nearby coastal cave complex known as Pinnacle Point, researchers found sixty pieces of hematite ore dating back about one hundred sixty thousand years. Hematite was processed to produce a red pigment and was used for functional purposes such as treating hides and in the making of adhesives for gluing tool parts together; it was probably also used to color tools, beads, and perhaps even bodies. It is not known if its use was symbolic at that early date, although at later sites, with definite evidence for modern humans, it was used symbolically. Further, Pinnacle Point does have evidence, thought to be symbolic, of a highly complex tool technology dating back seventy two thousand years ago. According to Tattersall, this technology required many stages of forward planning that would have necessitated the ability to abstract and visualize long chains of cause and effect.[87] Such evidence of symbolic reasoning, together with the more sophisticated jewelry and sculpture found at Blombos Cave, would certainly seem to indicate an accelerated form of cultural change occurring in Middle Stone Age Africa around seventy thousand years ago, at a time when the Neanderthals were the dominant species in Europe and where no such advanced artifacts have been found. At other South African sites, such as those at Still Bay and Howieson's Poort, sophisticated tools have also been found as well as sea shells and ostrich shells painted with red ochre in a way considered to denote symbolism; the latter industries span the millen-

[87] Ibid., pp. 200–202; see also Stringer, *Origin*, p. 127.

nia between sixty-five thousand and seventy-two thousand years ago.[88] Unfortunately, the Southern Middle Stone Age Africans who fashioned these tools and artifacts did not leave any fossil remains behind, and the paleoanthropological and archaeological experts can only assume, based on their correspondence with the later work of modern *Homo sapiens*, that they may have been the African equivalent of Cro-Magnons and hence the predecessors of those who later left Africa. Stringer is of the opinion that there may have been small pockets of behavioral moderns developing their culture and technology during that time, moving around from place to place. Significantly, the widespread use of tick-shell beads and red-ochre pigment throughout the range of anatomically modern *Homo sapiens* occupation, from South Africa to Morocco, between one hundred thousand and eighty thousand years ago—albeit without the complex social and symbolic connotations found at the above Middle Stone Age sites—may indicate that anatomically modern *Homo sapiens* were perhaps less culturally evolved than their European descendants but did already possess symbolic and artistic aptitudes. This would suggest that the distinction between anatomically and behaviorally modern humans may indeed have been cultural rather than cognitive. At any rate, at some momentous time between seventy thousand and one hundred thirty thousand years ago, our species emerged in its culturally modern form.

---

[88] Stringer, *Origin*, p. 49. See also Robert C. Berwick and Noam Chomsky, *Why Only Us: Language and Evolution* (Cambridge: MIT Press, 2016), p. 38. Berwick is a professor of computational linguistics and Chomsky is a world-renowned professor of linguistics.

The specific location of human origins may never be known. While the oldest fossil remains of our species have been found in East Africa, the signs of greatest behavioral modernity have come from the Middle Stone Age in South Africa. The continent is, of course, vast and remains largely unexplored; further fossil evidence will, inevitably, emerge to shed more light on our remote and mysterious African origins. Stringer has in the past favored the view that our species "evolved very rapidly in one small area—a sort of African 'Garden of Eden.'" Tattersall, on the other hand, tends toward the view that it was a punctuational, challenging environmental event of some kind that caused "one local African population of *Homo sapiens*" to emerge as fully symbolic from their trial and go on to populate Africa and the rest of the world.[89]

There is also the possibility put forward more recently by Robert C. Berwick and Noam Chomsky that it was a sudden punctuational development in language that caused our species to evolve as fully symbolic and distinguished it from all other hominid species.[90] They refer to language as "a curious biological object . . . which has appeared on earth quite recently. It is a species property of humans, a common endowment with no significant variation . . . unlike anything else known in the organic world in its essentials, and surely central to human life since its emergence."[91] Berwick and Chomsky assert that language is grounded in a universal grammar, the basic structure of which is innate in humans and is independent of sensory experience. The rudiments of universal grammar, they argue, are unlearnable

[89] Stringer, *Origin*, p. 202; Tattersall, *Masters*, p. 195.
[90] Berwick and Chomsky, *Why Only Us*, pp. 3, 64–66, 69.
[91] Ibid., p. 55.

without a prior innate knowledge of grammar. Despite the differences between the world's spoken languages, they all are based on this "universal grammar."[92] Berwick and Chomsky also refer to Wallace's conviction that language is a central component of "man's intellectual and moral nature."[93] Taking a reverse engineering approach, and basing their conclusion partly on the presupposition that we were and are the only species with language, they argue that for modern humans to have evolved language, the genetic leap that facilitated it could not have been gradualist and was due to a random mutation in perhaps just one individual. Hence language is perceived as a macromutation that produces a phenotype well outside the range of variation previously existing in the population. The scientific consensus is, of course, that other hominid species, including our own immediate predecessor species, did have some form of language and that this would give those species that possessed it a selective advantage. The strong likelihood, based on the evidence, is that there was a degree of gradualism in the evolution of language. Yet I am in agreement with Berwick and Chomsky in their conviction that human language is unique and somehow at odds with everything that went before. For the theist, the linguistic dissonance between human language and any other form of communication found in the natural world takes on another connotation. The universal structure of language as "a finite computational system yielding an infinity of expression,"[94] along with the fact that it is the central component of our abilities for symbolism, creativity of all kinds, and the ability to create intricate social

---

[92] Ibid., pp. 6, 58 and frequently.
[93] Ibid., p. 55.
[94] Ibid., p. 1.

structures, would seem to rule out a totally natural and gradualist evolution. Indeed, it resonates with Wallace's conviction that there was divine intervention in the evolutionary process; the attainment of language was undoubtedly a core element in the crystallization of our higher intellectual and moral faculties. Given that there is no satisfactory gradualist or punctuational explanation for how such an event might have occurred, Chomsky and Berwick, in highlighting the improbability of gradualist evolution for the capacity that separates us from all other species, could be said to be lending further weight to the argument that there was indeed divine intervention in relation to what is often called our uniquely "human capacity." This sudden leap in our language capabilities could have occurred either within the species, implying cognitive discontinuity between anatomically and behaviorally modern humans, or between our species and an earlier archaic predecessor one. Interestingly, Berwick and Chomsky argue that, due to its universality and the fact that the characteristic behaviorally modern skills associated with it appeared relatively recently in the fossil record, the full attainment of our capacity for language occurred fairly recently among a small group in East Africa [not long before the trek of modern humans from Africa] of whom we are all descendants, distinguishing contemporary humans sharply from other animals, with enormous consequences for the whole of the biological world.[95]

A qualitative distinction between anatomically and behaviorally modern humans may also be indicated by the fact that the former seemed able to coexist peacefully with Neanderthals, whereas the evidence is that the latter were not. It is also

[95] Ibid., p. 55.

possible however that the anatomically moderns remained relatively peaceful so long as they were still living in small groups prior to the attainment of full cultural modernity. Whatever the explanation, around sixty thousand years ago, one small group of fully modern humans migrated out of Africa on a great journey. Culturally, artistically, technologically, and cognitively they were vastly superior to any of their remaining archaic contemporaries; yet they would also turn out to be more predatory and warlike than any hominid species that had gone before. What factors could have brought about such a transformation? In searching for answers to this question, we will begin by drawing some important conclusions about the lifestyles of Neanderthal-type hominids, including anatomically modern *Homo sapiens*, who shared the planet with one another up to about one hundred thousand years ago. We will then imagine the likely impact modern intellectual abilities would have had on the not-yet-fully-humans who were our immediate predecessors.

The Moral Animal

It is reasonable to speculate that members of the *Homo* genus would, in general, have been temperamentally similar to bonobos and chimpanzees—playful, easygoing, affectionate, and hierarchical to a greater or lesser degree. It is also reasonable to assume that, due to the evolution of superior skills in hunting and food gathering, its members would have been even more cooperative than their first cousins in the natural world. They would certainly have been more sociable and affectionate for all of the reasons discussed: the control of fire for warmth, protection, and cooking would have led to social bonding, development in vocal communication, primitive forms of conviviality, and communal entertainment. The development of cognitive and technological skills far superior to those of apes would have made them empathetic to a degree beyond anything of which chimpanzee species are capable, as the fossil record shows; instinctive empathy would have evolved into the ability to give practical, long-term assistance to those who needed it. Above all, the combination of hominid cognitional capabilities with the arrival of pair-bonding and the nuclear family would have elevated male-female relationships and the relationships of fathers to their children far above anything of which any ape species was or is capable. As we have already remarked, there would also have been a marked reduction in male competitive aggression over resources and females. Moreover, we may reasonably speculate that it was during this period that the helpful and courteous behaviors that are the most attractive aspects of "reciprocal altruism" began to evolve. I would surmise that it was between the evolution of *Homo erectus* and the appearance of the first cognitionally modern humans that the basis for what we like to call common human decency was laid down. This fundamental decency remains one of the most ubiquitous and reliable characteristics of evolved *Homo sapiens*. The ability of different species of hominid to coexist with one another in the Levant for thousands of years provides further evidence of the peaceful nature of the *Homo* genus and, in my opinion, is an indication that this genus is indeed more closely related to bonobos than to chimpanzees. The evolution of increasingly attractive behavioral instincts and patterns would, of course, have provided a premoral basis for the qualitative increase in intellectual abilities that would establish the basis for the evolution of conscience.

We will now briefly recap the evolutionary explanations for the qualitative leap in intelligence that transformed a hominid species into a human one. Those evolutionary scientists who hold to the Darwinian view of evolution argue that it occurred through gradual, incremental evolutionary transformations. The fossil and archaeological records don't support this view, however, as the appearance of the full suite of human characteristics appears too suddenly to definitively establish a Darwinian explanation. The other viewpoint is that a punctuational event, such as that described by Berwick and Chomsky, transformed one hominid group with great suddenness into a new species. As discussed above, however, I would strongly argue at this point that a theistic explanation for the sudden acquisition of the basis for symbolic language must also be taken into consideration, and this would vindicate Wallace rather than Darwin. As you will recall, Darwin held to a difference only in degree between hominids and humans, whereas Wallace declared the difference to be qualitative. Wallace's profound conviction that evolutionary mechanisms could not have achieved the human intellect, grounded in logical scientific inference, led him to conclude that there must have been divine intervention in the evolutionary process. In explicitly religious terms, this would mean that God infused a spiritual soul into a preexisting hominid species. The latter is, in my opinion, by far the most likely explanation for what happened, and we will return to it in a later chapter in a theological context. It is also worth remarking at this point that Darwin's speculations on how conscience evolved are not dependent on natural selection having been its only catalyst, since, as he himself has stressed, the evolution of conscience would have necessitated a high degree of linguistic and intellectual ability, however such ability came into existence.

We are now in a position to speculate about the composition of prehuman societies prior to the attainment of full humanity.

It is reasonable to assume that the very earliest moderns would have existed in small, hunter-gatherer, fission-fusion communities; since modern humans are also able to live in large groups (something that chimpanzees cannot do without becoming violent), the small groupings may have been part of a larger parent population. The greatest distinguishing factor between all hominid and chimpanzee groupings would have been the familial structure of the former. We know that Neanderthals lived in small tribes composed of extended family, and while the evidence indicates that anatomically modern sapiens lived in much larger social groups, they would have been composed of nuclear families, with perhaps a mild level of polygamy among tribal leaders. The most striking similarity between chimpanzee and contemporary human society is, unfortunately, not the easy-going, playful affection of chimpanzee troops but that of male-dominance; indeed, we are chimpanzee-like in a peculiarly human way. Anyone who has seen the humanlike bad-tempered apes depicted in the *Planet of the Apes* films will understand what I mean. Human society, as in that of the common chimpanzee—albeit to a far more violent and dangerous extent—is controlled by coalitions of powerful males who, down through history, have cooperated in warfare and politics to monopolize the material resources of the planet. Such extreme levels of male dominance and violence would not have existed among the earliest human groupings, however, even if early humans were closer to chimpanzees than to bonobos. Due to pair-bonding, chimpanzee-like alpha-male territoriality would have been lessened rather than increased. If, however, the first humans were closer to bonobos than chimpanzees, they may even have been female-centered; at the very least, society would have been more egalitarian, with power being shared among coalitions of both males and females.[96] I would argue that societal structures may have been more bonobo than chimpanzee-like, with female and male coalitions, alpha females as well as alpha males, and a greater variety of types of leadership than obtains in evolved human societies. The latter would hold true, of course, even if early societies were more evenly balanced between chimpanzee and bonobo types.

Combining what is known about early hominid and chimpanzee societies, we can now speculate that the earliest almost cognitionally modern *Homo sapiens* would have been a vastly more intelligent combination of them all. Before speculating any further on what they would have been like, it must first be remarked that the initial leap in cognitive ability, however and whenever it occurred, would not have given them full intellectual equality with us. To paraphrase Darwin, they would at that point have barely or not quite achieved the rank of humanity. Interestingly, Darwin's view that the flowering of our language abilities occurred during a peaceful phase that was prior to but crucial to the attainment of our full cognitional abilities is perfectly reconcilable with the view of Berwick and Chomsky and, as we will later see, with the theological perspective on origins. For Darwin, this phase constituted a break in nature; for whether the acquirement of the structural basis of language was due to divine intervention, a punctuational event, or the result of gradual evolution by natural selection, there would still have to have been a transitionary period as language and its associated skills developed. The skill most crucial to intellectual modernity—and to moral decision-making—is, of

[96] De Waal, "Bonobo Sex and Society," p. 10.

course, linguistic, as Tattersall has pointed out in relation to the differences between hominids and humans where language is concerned:

> It is hardly overstating the case to characterize language as the ultimate symbolic activity, allowing as it does the generation of an infinite number of statements from a finite group of elements. Like thought, language involves dissecting the world around us into a huge vocabulary of symbols that are then combined, according to rules, to make statements not only about the world as it is directly perceived, but also as it might be. And it is virtually impossible to imagine our thought processes in its absence, for without the mediation of language those thought processes would be entirely intuitive and nondeclarative, merely involving the association of incoming stimuli with remembered ones, and responding accordingly. This is not to say that responses of that kind need necessarily be simple. Extremely complex associations may be made without requiring the process of abstraction that lies at the basis of symbolic thought. We know this from the example of earlier hominids. . . . But language facilitated the imposition of symbolic information-processing upon older cognitive processes. And this added an entirely new dimension to the way in which hominids saw the world, and eventually reimagined it.[97]

---

[97] Tattersall, *Masters*, pp. 214–5.

The arrival of symbolic capabilities would have marked a decisive rupture with the past but not the instant appearance of fully modern humans. There would have been a transitional period during which the reciprocal relationship between developing language capabilities and corresponding thought processes transformed the way these almost-humans perceived their world. During this time, there would also have been continuity with the past as transitioning human groups and individuals reinterpreted their previous cultural and behavioral practices under the impetus of developing external and internal symbolic and rational processes. As Darwin has pointed out, only at the culmination of this period would the ability for fully conscious moral decision-making have become possible, along with the beginning of our moral evolution.

We will now try to envisage the cultural situation of these barely human beings and to imagine what the immediate impact of their new intellectual capabilities might have been. All of the available evidence supports Darwin's speculations that their society would have been peaceful and family oriented, including his assertion that offspring would have been universally cherished and protected. There would also, undoubtedly, have been coalitions of males who cooperated in hunting and protecting their tribes and families, though not necessarily male dominance. If early societies were closer to chimpanzees than bonobos there may have been a degree of male dominance but it would have been to a far lesser extent than the type of dominance that has characterized human history. What Darwin, a typical Victorian male, could never have predicted is the emerging evidence for female coalitions of power and influence, and for alpha females as well as alpha males. It is also probable that the first humans were a species far more comfortable with their sexuality than we are, since they would have been guided by their habitual natural sexual rhythms and instincts. There would probably have been a greater homogeneity in levels of sexual interest and activity among adults than is the case among us, their more evolved descendants; in contrast, our own sexual behaviors range from the highly promiscuous to the prudish and from the debauched to the neurotically repressed.

I would argue that the initial impact of increased cognitional ability on these transitional humans would have been greatest on aesthetic sensibilities, the enjoyment of material pleasures and personal relationships. The environment would have been transformed into a place of beauty as well as function; the enjoyment of tribal life would have been greatly enhanced not just through higher aesthetic sensibilities but by developing powers of communication that would have increased the appreciation of shared experience. Developing language would also have contributed to the deepening of friendships, increasing intimacy in familial relationships, and the ability to empathize; due to the development of exclusive mate relationships and a more intense focus on child-rearing, loving and compassionate instincts and behaviors would have become conscious and hence more intimate. Last but not least, quantum leaps in hunting, cooking, and the ability to entertain through music, art, singing, and dancing would have hugely enhanced the capability for enjoyment that characterizes the chimpanzee and bonobo species. In summary, the evolution of the human family would in all probability have made these most mature of hominids and immature of humans an even more loveable and attractive species than chimpanzees, bonobos, or earlier hominids. Our earliest ancestors, the ones Darwin envisaged as far gentler than ourselves, may well have been peaceful, playful, technologically skilled, artistic, tolerant, egalitarian in comparison to ourselves, affectionate, or some other combination of characteristics that we can no longer properly imagine—a truly amazing species in relation to all others, one that had not yet learned how to be more predatory, more aggressive, and more treacherous than any other. Lest we get carried away by too idyllic a picture, however, there would also have been violent impulses and activities. As has already been pointed out, chimpanzee bands have been observed in violent territorial battles with other troops, and cannibalism seems to have been part of the hominid repertoire of behaviors; even in the most benign of species, there is always an element of "nature red in tooth and claw." Yet the social structures and milieu of these almost-humans had set the stage for the evolution of a particular type of conscience. The theologian Dennis J. Billy speculates that with the attainment of modern cognitional abilities, early humans would have begun their existence as a new species blessed with a strong communal self-awareness that would have given individuals "a sense of responsibility for and belonging to the larger whole."[98] In other words, this was a species that could have used its new intellectual and rational powers to lessen or overcome what violent instincts it possessed; equally importantly, these gentler, early human individuals could have chosen to balance the needs of the individual against the group in such a way that the overall good of the tribe would have taken precedence over short-term individual gain. This would have led to the evolution of altruism at group and species level, and a speedy (in evolutionary terms) detachment of human behavior from its biological roots. The human

[98] Dennis J. Billy, *Jesus, the New Adam* (Eugene: Wipf & Stock, 2017), pp. 6–7.

species would then have evolved through the type of cultural evolution described by neo-Lamarckists such as Gould to be at the very least a more humane, kindly, and good-tempered one overall than ourselves—not just intelligent and inventive, but potentially the most peaceful and cooperative one on the planet.

Tellingly, the archaeological record reveals that the anatomically modern humans who had coexisted peacefully with their hominid brothers for millennia changed rapidly into a warlike species of the genus *Homo* once cultural modernity had been achieved. It is known, for example, that once modern humans invaded Europe and western Asia about forty thousand years ago, the remaining *Homo erectus* in Asia and the Neanderthals in Europe—the last hominid species to go extinct—were gone within thirteen thousand years.[99] History provides abundant evidence of the horror inflicted on indigenous peoples by invading humans. How then might early humans have treated those who were not even of their species? As Tattersall and fellow paleoanthropologist Jeffrey Schwartz have pointed out, the speed of Neanderthal extinction in particular makes it reasonable to speculate that their destruction was probably due, at least in part, to direct conflict.[100] At worst, it could have resulted from a form of genocide. The latter possibility is made more likely by the poignant discovery of the last known Neanderthal refuge at Gorham's Cave in Gibraltar, where Neanderthals may have survived until as recently as twenty-four thousand years ago. Had they been forced down to the southernmost point of the continent when they were just at the point of extinction? At any rate we can deduce that human predatory and aggressive

[99] Ibid., p. 197.
[100] Ibid., p. 221.

characteristics must have evolved very quickly in evolutionary terms once the necessary linguistic abilities had been achieved. So what was the catalyst that precipitated such a dramatic change? In the following chapter, I will shed further light on the moral events of our primeval history.

Chapter 5
Paradise Lost: Alpha Males and Alpha Females

We will begin our initial speculations concerning early human moral evolution from a naturalistic perspective by imagining a time, several generations following the attainment of modern human levels of cognition, when language skills would have developed sufficiently for moral decision-making to become possible. By considering a period a century or two after the emergence of modern *Homo sapiens*, we are keeping our options open regarding beginnings, which may have been due to gradualism, a punctuated event, divine intervention, or a combination. We are also keeping our options open with regard to the Genesis account; in the next chapter we will focus on human origins from a more explicitly theological perspective. For now, however, we will concentrate on the situation that would have confronted our forebears during that primal period when new, more complex societal structures were beginning to evolve.

By that time, population density would have increased, and we will hypothesize that environmental conditions had deteriorated in some significant way. While we cannot be sure what the environment was like for the first generation of our species, paleoanthropologists such as Stringer have suggested that the first human grouping would have lived in a pleasant and fertile environment ("an African Garden of Eden") in which there may not yet have been any reason for fear, greed, envy, or treachery. One way or another, of course, it would have taken several generations for the first humans to develop the ability to plot and plan against one another. The question we must now address is this: How did the gentler species envisioned by Darwin and portrayed in the previous chapters give rise to the one that John Gray has described as the most predatory and dangerous on the planet? How, in a world that is both beautiful and fruitful, have we come within sight of annihilation and extinction through arms races and misuse of the environment? For over three million years, various hominid species hunted, gathered, made tools, and travelled together in ecological harmony with one another and with the natural world while evolving into ever more intelligent, cooperative, and empathetic creatures; yet the one hominid species that is still in existence engages in war, genocide, and environmental destruction in a manner and to a degree that is totally at odds with all that went before. As Harari points out, "We have the dubious distinction of being the deadliest species in the annals of biology."[101] Added to that is the poverty, inequality, violence, prejudice, and sheer hatred that scars a species in so many other ways remarkable in comparison with its hominid forebears. In attempting to answer the questions asked above, we will try to imagine the likely situations at the outset of human history that would have posed—for the very first time—conscious moral challenges to our earliest progenitors, starting with individual dilemmas.

[101] Harari, *Sapiens*, p. 82.

We will assume that, as Darwin maintained, tribal life was peaceful at first. Resources would have been distributed with relative fairness, and tribal members would have continued to cooperate with one another in the hunting of food and against threats either from the environment or dangerous animals. Affectionate social instincts and the ability to empathize, coupled with high levels of intelligence, would, however, sooner or later, have brought about the ability to obtain more than one's fair share, as well as knowledge of the hardship and suffering that could potentially be caused by greed or environmental threats. We will now try to envisage a situation in which a tribe, perhaps due to a disastrous weather event, is threatened with a serious shortage of food; its members are already on restricted rations and experiencing real hunger for the first time. The tribal hunters fan out to search for food and prey, and one of them has the good luck to find and kill a large bison in an area where there is also a plentiful supply of nutritious berries. His immediate instinct is to alert others of his tribe that he has found food, and he is just about to do so when he hesitates. A pang of fear causes him to speculate about his future well-being and that of his family. What if this is the only food that is found, and there will not be enough to last more than a few days? Perhaps he should keep his find a secret and save it all for his extended family, by which time they could have traveled farther afield and found more fertile ground. The hunter has never before suffered from hunger, nor seen the worst of its impact on others; he cannot fully know the devastation that a "selfish" option could cause. Let us suppose, in Darwinian fashion, that the powerful survival instinct, fueled by his own hunger pangs, causes him to suppress habitual cooperative instincts and choose himself and his family over the tribe as a whole. Having initially satisfied his own hunger, he conceals the rest of the meat and returns to his tribe. None of the other hunters has had any luck, and our hunter guiltily pretends that he too has been unlucky; for the first time in his life, he engages in deceit. Later, under cover of darkness, he and his family slip away when no one is watching. They struggle to reach the more fertile territory he has discovered and survive for a week or so, by which time the hunter unexpectedly comes upon a herd of bison and an area where fruit and vegetables are varied and plentiful. Overjoyed, he returns to the place where he left his fellow tribesmen and women only to find that many have died. Now he is in turmoil, knowing that if he had alerted the rest of the tribe as well as his family to the new territory he had discovered, they would in all likelihood have survived. He is filled with remorse and grief, horrified at the price paid for his self-interested decision by people he has been close to all of his life. He promises himself that should a similar situation occur sometime in the future, he would rather die alongside his tribal friends and neighbors than take the slightest risk of causing a disaster that could be avoided. Although our hunter may never be able to forgive himself for what has happened, he cannot be roundly condemned for his actions, since experience has not yet endowed him with the possibility to foresee all possible future contingencies and enable him to take a calculated risk. In this instance, newly evolved reasoning abilities, combined with fear and hunger, have led him to go against automatically cooperative and generous instincts. They have also led him into deceitful behavior in ensuring that his tribal companions were unaware of his good fortune in the hunt. Calculated strategies of deceit must have originated in such situations. In this case, however, there was no malicious or cunning intent, and the fundamental honest and affectionate nature of the hunter ensures that his action will not be repeated. The evolution of such strategies of deceit and their fixing in human nature would imply the repeated and deliberate suppression of generous instincts and the gratification of selfish ones by a significant number of the population, even when the ability to consistently foresee long term contingencies and consequences had developed. In my opinion, the evolution of cold-blooded ruthlessness and treachery would have been more a remote possibility than a probability.

One can also imagine similar situations, occurring during times of food shortages, when perhaps one successful hunting group in a tribe would reap the reward of a successful hunt by keeping all of the food for themselves or their families to the detriment of the rest of the tribe; alternatively, several individual hunters could decide independently but simultaneously not to share their supplies. At first, they would not have been able to fully envisage the resulting suffering and starvation of other members of their tribe, but they would have gradually become aware of the consequences of such behavior and would be forced to live with the memory. It is also likely that already existing selfish, greedy, or acquisitive impulses, combined with highly developed reasoning powers, could have precipitated such behaviors for the simple motive of self-aggrandizement in terms of material goods and status in the tribe; such behaviors were bound to be tried out by at least some individuals some of the time. Again, however, the detrimental results on still tender consciences, the affectionate social instincts, and the opprobrium of the tribe should have led either to the renunciation of such behaviors or their mitigation. There is also the fact that at such an early stage, when the majority of the tribe would still have been honest and empathetic, selfish and deceitful behaviors would have stood out; the socially skilled "conman" type would not yet have evolved.

The capacity to foresee long-term consequences would, as Darwin argues, be gained by trial and error over time; the relative fallibility of such judgements—especially in times of difficulty or crisis—would perhaps initially make it safer to trust in the relative infallibility of one's short-term interests. As we saw in the first example, the newly evolved trait of consciousness, the ability to recall at will, combined with the sensation of fear would have contributed to the suppression of a habitual instinct to share with the whole tribe. We can see, however, that a multiplication of such dilemmas continuing through time would have facilitated a much more comprehensive and nuanced degree of moral choice along with the development of tribal cultures that encouraged honesty, cooperation, and a fair distribution of resources.

I have made the above examples as dramatic as possible to illustrate how conscience would have developed. On a day-to-day basis, most moral challenges would have been more mundane—whether to share your food with others who had not been as successful in the hunt, whether to help neighbors who had difficulties of one kind or another, whether to impart new information that had been discovered, or to generously share with others new techniques you had invented for making tools or weapons. It is only to be expected that particularly hard-working or inventive individuals might have felt entitled to some extra reward or praise for their efforts or achievement and a certain degree—perhaps even a high degree—of competitiveness would always, in my opinion, have been essential to drive the species forward. We cannot be completely egalitarian. The problem is that the natural competitiveness of our species somehow attained a ruthless ferocity to the extent that a significant proportion of any human grouping is now capable of practiced deception, exploitation, cruelty, corruption, and even parasitism to attain wealth, power, and prestige. As I have argued above, this evolutionary pathway could only have been taken through deliberate, repeated, hardening of hearts that would have led to a change in tribal culture for, as Darwin repeatedly stresses, any individual who succumbed to selfish instincts that inflicted damage on his or her neighbor or the tribe as a whole would, sooner or later, have had to confront not only the disapproval of the tribe but also the persistent social instincts as they reemerged and made themselves felt. Even within the lifetime of the very first tribe or tribes, the powers of reason should have made it obvious that if everyone cooperated during times of both plenty and scarcity, the whole tribe would benefit in the longer term with regard to both health and happiness, even though such a state of affairs might occasionally and sometimes frequently have required individual sacrifice and risk. Humans, with their newly developed reasoning powers would have been able to devise cooperative strategies for efficient and fair utilization of the environment. Such strategies should have appealed far more to the affectionate, social nature of the first humans than coldly competitive ones. The combination of generosity, empathy, and increasing intellectual powers could have led to the achievement not just of vastly improved living conditions but to a further strengthening of the loving and affectionate social instincts. From our own evolved perspective, it would seem that the moral challenges I have described did not cause the first humans to overcome selfish instincts in the cause of the greater good; they must obviously have learned to practice complicated strategies of deceit and treachery in spite of a growing awareness of the consequences for their neighbors. This means that the genes of the harder-hearted and more deceitful (no matter how mild the deceit or slight the flaw in compassionate instincts) would have begun to increase incrementally and then to proliferate, since the more compassionate and honest would have been reluctant to indulge in violent or treacherous behaviors; hence they would have been less successful in material terms and in propagating their genes. As time went on, not only must our primeval forebears have frequently succumbed to their short-term needs at the expense of others, they must also have begun to calculate long-term selfish strategies. They learned, for example, to double-cross one another. There must have been many times when individual or groups of hunters would have stolen from one another, killed one another, or failed to warn one another of dangerous predators.

At this earliest stage of human existence, it would of course have been far easier to succeed at treachery than it is now. If most of one's family, friends, and neighbors were being cooperative and trusting, someone who was tempted to cheat would have had a clear field and easy pickings, so to speak. By the same token, the devastating results of treachery would have been impossible to overlook or play down, since counter-strategies would not yet have evolved. So, while those who cheated would have had a relatively easy time of it in one way, in another way it would have been far more difficult than it is now. There would be no escaping the consequences of one's acts because tribes would have been small and the suffering of friends and neighbors all too apparent. Moreover, in small tribes it would have been easy for the cooperative majority to devise sanctions against those tempted to cheat or betray. Added to this, feelings of remorse would have been far keener than before the affectionate social instincts had begun to weaken and when the level of societal competition and pressure had not attained anything like the level to which we are accustomed. Those who chose the paths of violence and treachery did so freely at a time when alternative courses of action would have been far easier to take than is the case in a world all too often determined by such strategies, one in which the stakes are much higher and temptation proportionately greater. Unfortunately, such would have been the rewards for their treachery that the genes of the treacherous, once their numbers had reached a certain threshold, would have spread with alacrity. There would also have been a knock-on effect in the evolution, for example, of suspicion and counterstrategies to anticipate and combat cheating. I have no doubt that Darwin is correct in his opinion that the greatest damage done to the original goodness of human nature lay in the diminution of affectionate instincts due to the calculated use of reason in the service of individual goals detrimental to the long-term interests of the group. Selfish and cruel actions can be executed by modern humans with barely a thought; this would not have been possible for our very earliest forebears.

We have so far speculated only on the corruption of the first tribal individuals. However, it is my belief that, given the basic good nature of the first human individuals, their lapses alone would not have been sufficiently frequent or serious to take evolutionary root to an extent that would seriously damage group and species morality. No man, as the saying goes, is an island: we are a social species and have evolved as individuals in relationship with various groups, both familial and societal. Most significantly, we are all individuals within a massively unequal pecking order, and, in my opinion, the root of real human evil lies in the evolution of the group strategies that have given rise, throughout human history, to cooperative male alliances excelling in the strategies of war and to the entrenched existence of powerful elites of one type or another. Although scenarios of the kind described above must have occurred some of the time, I believe that the main cause of the evolution of premeditated treachery lies in group rather than individual decisions. In what follows I will argue that the behavior of group leaders, in conjunction with the resulting interactions with individuals lower down the social scale, would have had a far greater impact on the evolution of human behavior than the conduct of the average individual.

Sexual Selection and the Morality of the Group

Warlike propensities are certainly a defining characteristic of our species and, as neo-Darwinists point out, warfare is an activity pursued almost exclusively by males acting in cooperative alliances headed by "alpha males." These alliances are deeply culturally embedded and have been the cause of war and poverty on an enormous scale. Evolved, complex human societies have always been shaped by alpha-male-led elites that are in overall control, and it is my belief that the evolution of treachery, greed, and violence would have been due in large part to the corruption of the earliest alpha males, since such corruption would have had a far greater impact on tribal morality than individual lapses. It is also my belief that—as with the average tribal individual—such corruption was neither inevitable nor probable. Why, then, might it have occurred?

Daniel Dennett argues in *Freedom Evolves* that evolutionarily stable strategies such as "lion and lamb" and "chimpanzees, bonobos and elephants" evolved from an original anarchic state in which everyone would have automatically betrayed everyone else when necessary in the battle for resources. Freud and Darwin also believed in a primal anarchy, caused by the evolution of permanent sexual activity in both the male and the female, which would have led to a significant diminishment of *Homo sapiens*' social manner of existence once full rational powers were achieved. Freud speculates further that the rampant and unmitigated violence unleashed by primal anarchy would have led to the first steps in civilization as we now know it. There must have been a re-forming of social structures that would have been harsh in relation to those that had existed before; to prevent unbridled individual aggression, the group would have had to appropriate for itself the power to use violence and deterrence against the individual.[102] From a theological point of view, of course, primal anarchy cannot have been inevitable since this would imply that God was responsible for it. In accordance with the methodology we have used in chapters 3 and 4, we will now speculate on the alternatives to anarchy permitted by the scientific evidence.

While I would agree with the speculations of Dennett, Darwin, and Freud that the first humans descended into a state of anarchy at an early stage, and with Darwin and Freud that its main cause was sexual aggression, I would argue that such a state was far from inevitable, since pair-bonding and hence strong monogamous tendencies would have been well established in our species long before the attainment of modern powers of reason. As we have seen, Darwin himself believed that such a state could not have prevailed at the very earliest period of modern

[102] Dennett, *Freedom Evolves*, pp. 141–167; Darwin, *Descent*, Vol. 2, pp. 358–359; Freud, *Civilization, Society and Religion* (London: Penguin, 1991), p. 184.

human history because men and women would still have been under the sway of instinctive habits. Yet evolutionary biologists maintain that male aggression and sexuality are strongly linked and that their coevolution was the cause of the evolution of warfare and indeed the whole patriarchal system in which men have been the dominant and women the oppressed sex. The anthropological evidence suggests that male aggression generally is always linked to competition over females. Throughout recorded history violent competition among groups of males has occurred over any scarce resource that affects the ability to attract and support women, with land being the obvious example. More directly, the victors in war and battle have traditionally viewed the capture of women as a desired benefit of their success.[103] Working backward from the present situation, we can logically assume that a main factor in the corruption of the earliest alpha males as well as in the formation of male coalitions in war and politics would have been the desire for access to as many women as possible and hence for land.

Yet in attempting to understand the origins of patriarchy and warfare, we are presented with a chicken-and-egg situation: Which evolved first, male power and control of resources, or female preferences for that power? For the male desire for power and resources could not have evolved if females did not favor the wealthier and more powerful males. Evolutionary psychologist David Buss has argued that the evolution of women's preferences for a permanent mate with resources would have required a vital precondition: "resources would have to be accruable,

---

[103] Donal Symons, *The Evolution of Human Sexuality* (New York: Oxford University Press, 1979), pp. 145–154; David Buss, *Evolutionary Psychology* (Boston: Allyn and Bacon, 1999), pp. 298–306.

defensible, and controllable by men during human evolutionary history." These conditions are certainly met in humans, since "territory and tools, to name just two resources, are acquired, defended, monopolized and controlled by men world-wide."[104] It seems probable that men's competitive strategies and women's mate choices coevolved, leading to a situation in which men could dominate in the domains of politics and resources. While male aggression and the female desire for resources were probably present even in the very first recognizably human tribe or tribes, the latter characteristics would have been mild in comparison with fully evolved tribes. As we saw in chapter 5, there is also the possibility that coalitions of females and the influence of alpha females would have had an inhibiting effect on male aggression; further, such female cohesiveness would have acted as a brake on any excessive demands for resources from individual females. How then did the desire for wealth, status, and prestige become such a major preoccupation of our species? This is a question that concerns relationships between the sexes and sexual rather than natural selection. To understand the sexual politics of male-female relationships and their relevance to the morality of the first tribes, we will now explore in more detail the effects of sexual selection on human behavior.

The Evolution of the Alpha Male and the Alpha Female

Sexual and natural selection are distinguished from one another in that the latter is a result of the differing abilities of individuals to adapt to and exploit their environments, while the former results specifically from variations in the abilities of individuals

[104] David Buss, *Evolutionary Psychology*, pp. 105, 203–204.

to acquire mates. Sexual selection is considered to be an auxiliary mechanism to natural selection since, in evolutionary terms, survival is more effective if the genes that facilitated it are passed on. It is driven, according to Darwin, by male competition for females and female discretion in their choice of mate. He identifies two types of sexual selection: the first, intrasexual selection, results from a competitive struggle among males "to drive away or kill their rivals" for females; the other is intersexual selection, which is based on female choice. Females choose as mates those males who are best able "to excite and charm" them.[105] The first type results in large body size, natural weapons of offence and defense (such as a stag's horns), and pugnacity; the second in the evolution of bright colors and ornamentation (such as a peacock's tail). Darwin also identifies a more passive type of selection operating on women; this is sexual selection for beauty, owing to the human male's preference for physically attractive females. Contemporary neo-Darwinists have expanded on Darwin's original assertion that sexual selection acts passively on women, arguing that female competition acting over thousands of generations has also caused the evolution of competitive behaviors among women, although the latter are characterized by verbal rather than physical aggression. We will return to female competitiveness at a later stage of the discussion.

Darwin notes that throughout the animal kingdom it is the male of each species that has been chiefly modified by sexual selection. He puts this down to male competition and female choice, and remarks that "the males of almost all animals [have] stronger passions than the females." He also speculates as to

[105] Darwin, *Descent,* Vol. 11, p. 398.

why the female, "with the rarest exception," is less eager than the male. The female, says Darwin, generally "requires to be courted" and is "coy." It is the male eagerness for the female, he argues, that gives rise to competitive behavior among males: "The 'law of battle' for the possession of the female appears to prevail throughout the whole great class of mammals. Most naturalists will admit that the greater size, strength, courage and pugnacity of the male . . . have all been acquired or modified through sexual selection."[106]

The second type of sexual selection, intersexual selection, is described by Darwin as "a more peaceful kind of contest, in which the males endeavor to excite or allure the females by various charms." Female choice of attractive mates has led to the development in many bird and animal species of elaborate and colorful ornamentations in the form of crests, tufts, mantles of hair, and brightly-colored plumages. Darwin usually cites the courtship behavior of birds when illustrating the male power to charm: "Many male birds do not so much pursue the female, as display their plumage, perform strange antics, and pour forth their song in her presence."[107] In *The Descent of Man*, Darwin ponders on the evolutionary origin of the human love of music and the capacity to sing, which is common to almost all human individuals; he surmises that these abilities must have resulted from sexual rather than natural selection, since music has no functional role, and musical sounds and singing are used by many other species during their breeding seasons:

---

[106] Ibid., Vol. 1, pp. 272–274; Vol. 11, pp. 312–313.

[107] Ibid.; see also Vol. 1, p. 272.

> With all those animals, namely insects, amphibians, and birds, the males of which during the season of courtship incessantly produce musical notes or mere rhythmical sounds, we must believe that the females are able to appreciate them, and are thus excited or charmed; otherwise the incessant efforts of the males and the complex structures often possessed exclusively by them would be useless.[108]

In the previous chapter we speculated as to whether hominids would have been able to sing. While none of the species of Great Ape has such a capability, it is an interesting fact that the monogamous gibbons have singing voices that are said to rival those of operatic sopranos and even perform perfectly coordinated duets during their mating season. Of further interest, perhaps, is Franz de Waal's remark that when he first heard bonobos vocalize, "their high-pitched, melodious calls" reminded him of gibbons.[109] According to Stringer, Neanderthals would certainly have had the vocal capacity to produce musical notes, and given our own musical abilities as a species, it is almost a given that our ancestral species would have been able to sing. Darwin believes that these faculties must have appeared at a very remote period in "the half-human progenitors of man," and perhaps musical powers were another feature that distinguished hominids from their ape cousins. With regard to early humans, Darwin conjures up a delightful image of prehuman males and females singing to one another during the season of courtship, before the

[108] Ibid., Vol. 11, p. 333.

[109] Franz de Waal, "Behavioral Contrasts Between Bonobo and Chimpanzee", in *Understanding Chimpanzees*, p. 154.

power of articulate language had been acquired. Such behaviors would have continued through to early humanity, but must, he says, have belonged to an extremely remote period "before the progenitors of man had become sufficiently human to treat and value their women merely as useful slaves."[110] Darwin suspects that singing would have been part of the armory of the male "power to charm" and hence more effective in male courtship of females than vice versa; he admitted that his musings on the subject were speculative, however, and based on analogy with sexual selection among animals and birds. This was no doubt due to the fact that Darwin did not see music being used in this way in his own inhibited Victorian culture. For those of us, however, who are accustomed to globalization, the mass media, and the consequent ubiquity of "celebrity" male singers of various types, there is nothing remote or mysterious about the male "power to charm" the female through song. Judging by the female reaction, this power is incalculably greater than Darwin thought; he would no doubt feel gratified at such vindication were he still alive to experience it. We will consider the implications of this reemergence of the peaceful type of sexual selection in a later part of the chapter.

With regard to the phenomena of male eagerness and female choosiness, Darwin and his neo-Darwinist successors explain them in terms of the marked differences in the minimum contribution each sex must make to child-rearing. This difference basically boils down to the fact that males are capable of begetting large numbers of offspring in a short space of time if they fertilize a variety of females. A female, on the other hand, has a limited number of ova that may be fertilized and must endow

[110] Ibid., pp. 336-7.

each new embryo with all of its nourishment by incubating it in her body for nine months, followed by years of intensive child-rearing.[111] Even the most prolifically fertile woman can only produce a small fraction of the number of offspring that a prolifically fertile man is capable of producing. One of the sorest points in the perennial battle of the sexes is that men can avoid or refuse to look after their children, abandoning them—even denying paternity up until the very recent past—and leaving the mothers to do all the work and take all the responsibility. During evolutionary history, our female ancestors would have been seeking mates who were not only "charming" but also able hunters and reliable providers. For this reason, there would always have been a closer association between natural selection and sexual selection in men than in women, since the male ability to provide for and protect his family would have been an important factor in a woman's choice of mate.

The behavior of some males notwithstanding, the level of human male input into parenthood is unprecedented. In contrast to other primate species—and indeed mammals generally—the human male generally lives with his wife and children and expends a great deal of time and effort in looking after their welfare. In return for the constant availability of sex and to ensure the survival of their vulnerable young, male hominids would have lived with their mates and helped with the care of offspring. Our species is, however, still classed as polygamous, albeit mildly so in relation to other primate species due to the level of time and effort men are willing to devote to

---

[111] Richard Dawkins, *The Selfish Gene*, pp. 141–142.

parenting.[112] While de facto monogamy is generally practiced by humans and is held up as the ideal in many cultures, the vast majority of human societies are classified as polygamous because at least some men have more than one wife. In societies where polygamy is illegal, informal polygamy is clearly manifest in the ability of many wealthy and high-status men to attract multiple sexual partners and to practice serial monogamy by replacing older wives with younger "trophy" ones. While the human female is considered to be more selective in her choice of mate than the male, however, various biological factors indicate that women are not naturally monogamous. Contemporary neo-Darwinists suggest that, in comparison with our closest female primate relatives, women fall somewhere between the promiscuous chimpanzee and the monogamous gorilla. Overall, however, evolutionary biologists consider the main distinguishing feature between male and female sexual psychology in the human as well as most other animal species to be the male desire for a variety of sexual partners as opposed to the female desire for quality of survival abilities and commitment. Total monogamy would have been unlikely to evolve in our species, and the contrasting interests of men and women enhance interest and chemistry between the sexes; moreover, since these opposing drives are resolved in the human species through the phenomenon of "falling in love" and the desire for intimacy, commitment, and a fulfilling family life, they also link the amoral qualities of "fitness" with morality and virtue in a challenging and potentially fulfilling way.

---

[112] See Donal Symons, *The Evolution of Human Sexuality*. See also Christopher P. Sloan, "The Origin of Childhood," in *National Geographic*, November 2006, pp. 148–159.

In all cultures, those men who do have multiple wives and sexual partners tend to be those who have the wealth to support such a lifestyle. Hence human alpha males will be those who possess wealth, status and power, while human alpha females will be those who attract the highest status males. In evolutionary terms, the alphas of both sexes would have reproduced the most prolifically and successfully, thus passing on more of their genes. In an imitative species such as ours, they would also have had the most influence on cultural behaviors, inspiring others to emulate them. This coevolution of genes and culture explains the entrenched human desire for and admiration of wealth, status, and prestige. The depth and extent of such desires has given rise, as remarked above, to ruthless competitiveness where the acquisition of wealth is concerned, to violence and to inequalities in wealth and poverty. Given the close association identified from the anthropological record between warfare and competition over females,[113] we may conclude that the descent into anarchy was indeed due to sexual rather than natural selection. This means that of the two types of sexual selection operating on males identified by Darwin, it was the aggressive, competitive type that came to predominate in the human species. Yet, as we have speculated, the moral distance traveled between the behavior of the very first humans and those who descended into anarchy is in all likelihood too great to have been inevitable; I would argue that, based on the arguments presented in chapters 3 and 4, the peaceful, "charming" type of sexual selection should have been more likely to gain the ascendancy in human behavior, or at the very least to have maintained an equilibrium with

---

[113] See Symons, *Evolution of Human Sexuality*, pp. 145–154; Buss, *Evolutionary Psychology*, pp. 298–306; Darwin, *Descent*, Vol. 2, p. 323.

the aggressive type, hence holding it in check. Let us remind ourselves again of the earliest human groupings we imagined at the end of the previous chapter, to that rather idyllic image we can now add the pleasurable and exciting rituals of courtship, including musical ones. Darwin was convinced that sexual selection's most attractive effects would have been accomplished just prior to and including this time, before the powers of reason had begun to take their toll. During that remote period, it would have been the charming rather than the aggressive kind that predominated, characterized mainly by music, singing, personal charm, perhaps even an early form of oral poetry once language had developed to a sufficient degree, and an emphasis on physical appearance in both sexes. More importantly, as Darwin surmised, women would not yet have been as preoccupied with wealth and social status as are their evolved descendants, and so there would not have been so much male competitive aggression over the accumulation of resources.[114]

While this more gentle and beautiful type of sexual selection still exists, it is not as pervasive as it might have been, though it can be highly evolved at the individual level. At group and species level, the aggressive, power-seeking type prevails. Interestingly, the recent #MeToo movement is highlighting this phenomenon to an unprecedented degree. To imagine the impact that the peaceful type of sexual selection might have had if it operated at cultural level, we need look no further than books such as Malory's *Le Morte d'Arthur* or Tennyson's *Idylls of the King*, which narrate the legendary tales of King Arthur and the Knights of the Round Table, along with their associated tradition of chivalry. The round table at which the king and his

---

[114] Ibid., p. 334, p. 368.

knights sat was designed to foster equality and harmony instead of hierarchical status and self-importance. Serious disagreements were solved, where possible, through jousting competitions rather than war, and knights dedicated themselves to truth, honor, and the protection of the vulnerable, especially women and children. Women themselves were to be treated according to the highest standards of the knightly code of chivalry. While it is impossible to know whether or not there is any historical truth in these mythologized accounts, we can certainly draw on two transient examples whose historical credentials are not in doubt. The first is the medieval tradition of courtly love—related to the tradition and tales of chivalry—which inspired troubadours and poets to write songs and poems for unattainable noblewomen whom they loved chastely from afar. Interestingly, the latter tradition emerged in response to the brutality and crudity of the feudal system, which saw aristocratic women used as dynastic pawns in arranged marriages between the noble houses of Europe. The second instance is more recent and has already been cast in an evolutionary context. It has often been said by primatologists that the "make love not war" generation of the 1960s was a group manifestation of bonobo-like behavior. It was a decade characterized by high ideals, a powerful desire for peace, and the challenging of authoritarian conventions. Anyone who was alive in that golden decade will never forget the courage and eloquence of those such as U.S. President John Fitzgerald Kennedy, his brother U.S. Senator Robert Kennedy, and the Rev. Martin Luther King Jr., each of whom paid the ultimate price for their ideals. Interestingly, the Kennedy administration is often referred to as "Camelot."

The backdrop and also, to some degree, the catalyst for this "soft revolution" was the huge popularity of protest folk and rock music and the bands associated with it. The scenes of female mass hysteria that came to characterize concerts given by the Beatles and the Rolling Stones, among others, showed the full impact of the male "power to charm" as opposed to the male power to dominate. As with the courtly love of the medieval period, the sixties culture arose in response to barbarity, albeit on a hitherto unprecedented scale, in the savagery unleashed by World War II. Although such cultural manifestations tend to fizzle out, they present us with a tantalizing glimpse of what might have been; the hedonistic excesses of the sixties aside, its peaceful ideals and creativity reveal both the possibility of, and the powerful desire for, alternative, gentler, more artistic modes of existence. Interestingly, the female mass hysteria displayed at sixties rock concerts suggests a release of tensions that indicates repression. There may very well be powerful cultural forces as well as genetic factors at play in the triumph of the aggressive over the peaceful type of sexual selection at societal level, which would imply that cultural adjustments might enable previously culturally repressed more charming traits to reemerge. That, however, is a topic for another book.

What types of humans might have evolved had the peaceful type of sexual selection prevailed? It would certainly have produced gentler, more sensitive if no less courageous men and women, whose main concern would have been the development of character, personal charm, and courtship skills as opposed to the acquisition of wealth and social position. Men would have tried to charm rather than dominate women, which would have led to greater harmony both between and among the sexes. If, as Darwin surmised, women, as well as men, had chosen mates primarily for beauty and the power to charm rather than status at that earliest stage, the average level of good looks and general attractiveness to the opposite sex in both men and women would probably have increased. The world of work would have evolved to be fairer and less amorally competitive, more concerned with professional pride and job satisfaction than with the acquisition of wealth; in such a world, enjoyment would have taken precedence over status. Most of all, there would not have been any need for war. Competition would not have increased to the level that it has, and a person's status would therefore have related more to moral caliber than material success, to intelligence, artistry, creativity, craft, athletic prowess, and courtship skills rather than power or wealth. In such a case, both men and women would, in all likelihood, have been generally satisfied with sufficient material resources for comfort. There would have been no serious inequality with regard to wealth and nor would there have evolved the characteristics of ruthless ambition, aggression, cunning, and emotional coldness that result from its pursuit. That is not necessarily to say that there would not have been wealthy people or different strata in society, but that there would not have been extremes of wealth and corresponding poverty. We find ourselves, however, in a world marked by such inequality, in which power, wealth, status, and prestige are presented as the most desirable human goods. It is my opinion that a large part of the answer to what went wrong at the outset of our history lies in finding out why aggression rather than charm prevailed in the evolution of the human male. I have previously argued that this situation has resulted from an event or events at the outset of our history that caused the corruption of the alpha males and females. In trying to determine what actually happened at that crucial period, it will help to further clarify what exactly is meant by those terms as they are presently understood; the original leaders, both male and female, would have had different roles and characteristics.

The alpha males who concern us here are typically those who head up male coalitions and wield power in business, politics, and societal institutions generally. Although women in the Western world have made much progress in business and politics, and the number of females who occupy powerful positions in the business and political spheres has increased significantly, they remain a small minority in relation to their male equivalents. Such women are not "alpha females" in the traditional sense, however: up until the very recent past, the alpha female has been one who gains wealth and influence through her position as either the wife or the mistress of a powerful man. It is an undeniable fact that, throughout history, the most powerful women have been those who have leapfrogged to the top of dominance hierarchies and exerted influence over them through attracting the most powerful men. I would go so far as to argue that the ultimate alpha female throughout history has been the royal or political mistress, one who attains power mainly through strategic use of her sexuality. The scheming adulteress is an archetypal figure for whom there is no male equivalent, and in trying to envisage the corrupting or treacherous events of human prehistory, the temptation is to imagine a scenario in which a scheming, primal female seduces a susceptible tribal leader in order to increase her resources and influence within the tribe and who then urges her conquest into an abuse of his authority so that he will enrich himself for her benefit. The problem with such a scenario is that it couldn't have occurred, since the "scheming adulteress" type would not then have existed. A clue to the primal corruption of alpha males and females may lie in features of evolved tribal life remarked on by Darwin in relation to sexual selection:

> At present the chiefs of nearly every tribe throughout the world succeed in obtaining more than one wife. Until recently . . . almost every girl in New Zealand, who was pretty, or promised to be pretty, was *tapu* to some chief. With the Kafirs, "the chiefs generally have the pick of the women for many miles around, and are most persevering in establishing or confirming their privilege."[115]

He also points out that a weak man in a tribe, unless he is a good hunter and well liked, will seldom be allowed to keep a wife who a stronger man thinks deserves his attentions.[116] In other words, through the effects of the "the law of battle,"—the aggressive type of sexual selection—the richest and most powerful men could, in primitive evolved tribes, take whichever women they desired, hence bypassing the efforts of courtship. At an early stage of our primeval past, it would seem that pugnacity as a mating strategy was favored over charm and gained an evolutionary advantage. Why? We will shortly consider some of the situations that could have given rise to the triumph of aggressive power in sexual selection over the gentler, though far more attractive, powers of courtship and explore further how that would have played out in the evolution of the sexes. In seeking to understand what actually happened, I have found

---

[115] Ibid., p. 369. See also p. 363.
[116] Ibid., p. 324.

evolutionary biologist E.O. Wilson's concept of hypertrophy to be extremely useful.

The End of Innocence

Wilson has written that in his opinion "the key to the emergence of civilization is hypertrophy, the extreme growth of preexisting structures. . . . The basic social responses of the hunter-gatherers have metamorphosed from relatively modest environmental adaptations into unexpectedly elaborate, even monstrous forms in more advanced societies."[117] So, for example, the tendency in hunter-gatherer societies for one or two of the most successful men in the tribe to have more than one sexual partner during the course of their lives is transformed, in the context of certain "civilizations," into the phenomenon of harems:

> Harems constitute the hypertrophied manifestations of male ambitions released from the usual constraints of limited personal power, the fantastic products of a male psyche that evolved in social milieus in which extreme polygyny was impossible.[118]

The most prolifically fertile male in recent history is thought to have been Genghis Khan, whose "habit of impregnating large numbers of women in conquered populations" is legendary,

---

[117] E. O. Wilson, *On Human Nature* (Cambridge: Harvard University Press, 1978), p. 89.

[118] Margo Wilson and Martin Daly, "The Man Who Mistook His Wife for a Chattel," in *The Adapted Mind: Evolutionary Psychology and the Generation of Culture* (New York: Oxford University Press, 1992), Jerome H. Barkow, Leda Cosmides, John Tooby, eds., p. 301.

as is the historically documented fertility of his male descendants.[119] In parallel fashion, simple tribalism is transmuted, in a global context, into nationalism, racism, and religious intolerance. The tribal pecking order that would permit a few of its members to amass significantly more resources than the others becomes, in the civilized world, the blueprint for vast inequalities in wealth and poverty, while the wars that took place among our hunter-gatherer forebears over land and women are transposed, in the world of so-called civilization, into "sophisticated military organizations and techniques of battle . . . [that] in our time have . . . [brought us] to within one step of nuclear annihilation."[120]

I would argue, therefore, that the decisive event or events in the evolution of patriarchy and warfare might seem a small matter to contemporary eyes regarding the material form that it took, albeit that it has had devastating consequences; in the context of the earliest human grouping, however, it would have constituted, at the very least, a serious breach of tribal trust and cohesion. While we can never know for sure exactly what happened, the rapid transformation of the gentle species we have envisaged into the warlike one that erupted into Eastern and Western Europe some forty-five thousand years later implies a cultural change that came from the top down. A significant change in cultural norms driven by the tribal leadership would have had a much faster and more comprehensive impact on behavioral evolution than the piecemeal effects of unrelated individual lapses by ordinary tribal members. Further, given the imitative nature of our species, the human need for leadership

---

[119] Stringer, *Origin*, p. 177.
[120] E. O. Wilson, *On Human Nature*, p. 117.

and the fact that human nature was new, fluid, and untarnished, the behavior of the earliest leaders in the earliest societies would have had a powerful impact on cultural moral evolution and hence consequent genetic moral evolution, for good or ill. In light of what Darwin has to say about the crude tendency of the leaders of evolved tribes to take multiple wives by virtue of their superior power and potential to use force, we might assume that the first tribal leader used corrupt or forcible means to gain extra resources and hence the capability to acquire extra wives or other men's wives. Such brutal behavior, however, would not then have evolved any more than would its scheming, female counterpart.

Given the inextricable link between the evolution of warfare and relationships between the sexes, and the fact that neither the marauding chief nor the wily seductress types would then have existed, one possibility is that the original, primal injustice was perpetrated by the first leader with the consent or collusion of his wife; there was no initial chicken-and-egg situation as suggested by evolutionary biologists. Whether the first leader and his wife headed up a single tribe or an aggregation of tribes, the results would have been the same. Since the evidence thus far indicates that the very first grouping of humans who were both anatomically and behaviorally modern consisted of a single, small population or tribe, we will—for the sake of convenience as much as accuracy—begin our further speculations on the primal moral events of prehistory with one group in mind.

At the very beginning, as suggested above, it is most likely that tribal members would have divided hunting spoils and other resources with relative fairness; the chief or important female leaders in the tribe might have intervened if a dispute arose, and the most influential members of the tribe might even have taken a slightly greater share for themselves. With the acquisition of full moral powers there were, broadly speaking, two main courses of action that could have then been taken. The tribal leader could have encouraged full cooperation in the cause of the overall welfare of the tribe, leading to maximum and just use of environmental resources; such a course of action at the very beginning could have led eventually to the type of group and species selection described by neo-Lamarckism. Alternatively, he could have decided to enrich himself at the expense of the rest of the tribe. In other words, the choice was to become less or more acquisitive, to use power in a just way or to abuse it. Tribal chiefs who were fair and just themselves would no doubt have done their best to encourage and oversee the good behavior of the rest of the tribe. By the same token, members of a tribe that was ruled justly would have tended to follow the example of their leader and would generally have been content to maintain the status quo. A tribe whose leader behaved with honesty and integrity would have been one with little or no grounds for jealousy, envy, or covetousness and, through example, would have inspired similar behavior among tribal members. It is my opinion that the possibility of tribal leaders deciding to take advantage of their position to acquire wealth would, even under optimum moral conditions, always have been a possibility with the passage of sufficient time and would undoubtedly have occurred some of the time, though not on a sufficiently regular basis (or scale) to affect the overall moral evolution of the species. At the very outset, however, such a course of action seems highly unlikely because, at first, regular patterns of behavior would have been adhered to in matters related to survival as well as to reproduction. As I have previously argued, the transformation of a peaceful into a warlike species must have been the result of a significant cultural change that occurred in the very first human grouping and that could only have come from the top down. We will speculate, therefore, that the first couple agreed on a course of action that was a betrayal, at the very least, of tribal cultural mores up to that time. At this point in our discussion, we will focus on the corruption of the first tribal leader in a naturalistic way. As I will later argue however, such putative behavior in all likelihood would have resulted from a prior spiritual misdemeanor.

Understood in a naturalistic way, the serpent in the paradisiacal situation may have been the desire of the first couple, at some point, during those first years or decades, to "lord it" over the rest of the tribe by keeping for themselves a greater proportion of tribal resources than would have obtained previously; such behavior would, at the very least, have caused resentment. Meanwhile, those most compliant and loyal to their leader in the changed situation may have been slightly preferred in terms of resources and status over the other tribal members. During the first generation, this may very well have been all that occurred. The subtle shift in emphasis from just leadership intent on fair distribution of resources to a more power-based one focused on wealth and status, together with the beginnings of an elitist class would, however, have amounted to a change in culture that was carried over into successive generations and exponentially strengthened. It can easily be imagined how, with the passage of time, other men and women would have imitated the behavior of their leaders and the burgeoning elite. Simmering resentments among the nonconforming or nonpreferred tribal members would also have increased. Thus, the tribal culture would have been transformed from being cooperative and fair—even if hierarchical—to a divided one that overall was becoming ever more preoccupied with status, wealth, and the benefits of wielding power.

Eventually a chicken-and-egg scenario would have developed in which previously mild polygamous and promiscuous tendencies were strengthened in both sexes. As evolutionary biologists point out, the exacerbating of the need to compete aggressively for females would, among males, have led to selection for higher sex drives that motivated them to engage in the kinds of behavior that would enable gratification. By the same token, tribal women would have emerged who preferred wealth and power over "charm" and who were prepared to use their sexuality as a means of gaining them, in treacherous as well as legitimate ways. Competition among men to attract such women would have increased, leading to violence among men and growing inequality; meanwhile, heightened competition for wealthy mates would have caused an increase in female competitiveness, in the development of calculated and treacherous strategies of seduction, and hence in female sex drives also.[121] The evolution of competitive aggression and strong sex drives would thus have been simultaneous. Most crucially for the evolutionary process, the application of human rational powers by both sexes in the devising of treacherous strategies would eventually have resulted in the fixing of such strategies in the evolving population. Tribal leaders would also have used the social force of power rather than the more time-consuming rituals of courtship to persuade the desired female to choose him rather than his competitors. Males lower down the expanding tribal hierarchy would have become envious, and competitive jockeying for position would

---

[121] See Randy Thornhill and Craig T. Palmer, *A Natural History of Rape* (Cambridge: The MIT Press, 2000), pp. 53–54.

have stepped up. Once it became obvious that the most successful males in the tribe in terms of resources could quickly acquire multiple mates, the level of competition over resources would have escalated in ratchet-like fashion, with the satisfaction of short-term impulses taking precedence over the long-term welfare of the tribe. Further, if tribal leaders were focused on self-aggrandizement, they would not have given the kind of example that encouraged kindness and cooperation, nor would they have had the time or inclination to ensure that justice prevailed among their people. Inevitably, the anarchic situation described by Dennett would have prevailed, resulting in the establishment by natural selection of new evolutionarily stable strategies, or ESSs, to restore balance among the various evolving and oscillating traits. While this would have had the effect of limiting aggression, it would also have limited altruism; it would, moreover—due to evolved rational powers—have divided the human species into predators and prey. Tragically, the peaceful way of life that had continued over untold millions of years of hominid evolution would then have become a thing of the past. As Freud has argued, since this was a sentient and intelligent population, there would also have been major cultural change in an attempt to restore order. For Freud, this was the first major step in the creation of civilization as we know it, and it developed at an enormous price: the individual is pitted against the group resulting, all too often, in mutual hostility. Freud envisages a situation in which sexual and acquisitive instincts had been so strengthened by natural selection that the individuals who were physically strongest and most aggressive would have broken through previous societal norms and structures to run rampant, taking women and resources by force and leaving murder and mayhem in their wake. Sooner or later, a majority of weaker individuals would have come together that "is stronger than any separate individual and which remains united against all separate individuals."[122] This would have involved the imposition of force and the constant threat of force on populations in a new kind of ruling order, and the resulting resentments would have set the scene for future rebellions. To complicate things further, governing groups would have been more susceptible to subsequent corruption than the individuals under their control, since the original catalyst for upheaval had been the behavior of the first leader and his allies. By then, the desire for wealth and status, along with the traits necessary to acquire them, would have become culturally and genetically embedded. Inevitably, governing coalitions would have transformed into elites, and there would have been more rebellions, as other groups came together to seize the apparatus of control and hence of resources.

As population levels increased, it would have begun to occur to those who wished to accumulate wealth, power, and influence that to take the risk of attacking other tribes to steal their resources might bring great rewards. This would have led to further evolution of cooperative male strategies in warfare and politics, to the subjection of the innocent, especially children, to unnecessary risk, and, inevitably, to vastly increased levels of violence and bloodshed. It would also have led to the treatment of women as just another resource to be won or lost in battle rather than to be courted; by that stage, domination rather than charm would have come to characterize male-female relationships. Leaders endowed with the ability to successfully deceive would also have evolved. Since ordinary people, then as now,

---

[122] Freud, *Civilization, Society and Religion*, p. 284. Darwin, *Descent*, pp. 358–359, p. 367.

would presumably have wanted a peaceful existence and an equitable distribution of resources, leaders would have had to be able to give the appearance of justice even while initiating conflict and feathering their own nests. History has thrown up many examples of powerful men who have been able to fool the masses with a mixture of charisma, charm, and deviousness. It has also thrown up dictators who combine deviousness with a megalomaniacal disposition. Indeed, the masses are often grateful to be led by a benign and capable alpha male and to tolerate excesses where wealth and women are concerned. This is not to say that leaders are always corrupt in one way or another; there have been many exemplary leaders throughout history. The systems that produce them, however, easily tend toward corruption.

Sexual Selection: The Triumph of Aggression

At the beginning of the chapter we tried to visualize the moral transgressions of which the very first individual humans would have been capable, and I argued that the more serious ones I described were unlikely to have arisen at the individual level. In contrast, the impact on individual behavior of the type of cultural transformation described above would certainly have caused the rapid evolution of cold, calculating, and treacherous strategies along with the traits necessary to employ them. Based on the evidence, we can say with some degree of certainty that the primal moral events of prehistory must have involved leadership at the male and female level due to its impact on sexual as well as aggressive and acquisitive traits and the inextricable relationship between these traits. This impact has not just affected individuals; it has increased, as is particularly evidenced in its hypertrophied manifestations in complex societies, whatever hierarchical, acquisitive tendencies our progenitor species may have had. In the earliest human and prehuman societies, chiefs might have had at most one extra wife; while males lower down in the tribal pecking might have sought to emulate him, the difference between the sexual and economic success of men in any one tribe would have been relatively small. We now live in a world that has produced men as prolifically fertile and powerful as the aforementioned Genghis Khan, and consequently in which men are competing and striving in a pecking order whose direction and scale would have been unimaginable in even the earliest modern human populations, let alone among our prehuman forebears. In parallel fashion, the wealth and status toward which female competition is directed is fabulously beyond anything that could have been achieved or imagined by our earliest female ancestors.

It is my opinion that many of the differences in sexual psychology between men and women described above would have in some way characterized the sexes even had the charming type of sexual selection shaped the course of evolution. It may be concluded, however, that such a vastly increased scale and level of competition has had a deleterious effect on interpersonal and societal relationships that include relationships among as well as between the sexes. I would go so far as to suggest that the most successful survival and sexual strategies of men and women have stamped each sex with its defining characteristics. The male sex has been shaped by the need to cooperate in warfare, an activity pursued exclusively by men acting in cooperative alliances. Many evolutionary biologists have asserted that selection for skills in cooperative male violence has played a significant role in the evolution of human intelligence itself.[123]

[123] Buss, *Evolutionary Psychology*, p. 145; pp. 298–306.

War is an intensely cooperative venture, and male competitive aggression has given rise to an extraordinary ability among men to cooperate and form coalitions and alliances on the battlefield and in the political arena. Moreover, in any field in which many men work together—certain religious domains and the domain of finance immediately spring to mind—the pecking orders, hierarchies, and tactics that characterize male behavior within large groups will also manifest themselves, including aggressive jockeying for position within the pecking order. Indeed, the male sexual persona overall has become far more closely associated with the male work persona than might have been the case. One can barely imagine the result if the cohesiveness of the male sex had evolved through fair competition for both females and resources, with the consequent prioritizing of justice over competition.

With regard to the female sex, the harm is also great; for it seems highly likely that women have lost a great deal of their cohesiveness as a sex. The development of individual techniques of seduction in the pursuit of wealth and its associated power and status were the most successful ones for women, albeit that they would at times have involved putting one's own interests before those of the other women of the tribe and sometimes the betrayal of other women rather than cooperation with them as heretofore, a cooperation that had endured for millions of years. I have assumed that the first tribal leader would have been male, even if corresponding alpha females and coalitions of influential females also existed who were politically influential. This is because, due to the superior physical strength of men and the dependency of vulnerable infants on their mothers, hominid males would have been the overall protectors of tribal women and children and hence the most politically engaged.

Such male coalitions could also, of course, have cooperated to oppress females through brute force if they chose, an option that couldn't operate in reverse. Female coalitions would, of necessity, have been centered around cooperation in child-rearing and the gathering of food. With the evolution of modern intellectual capacities, the emergence of complex civilizations, and in a situation in which the peaceful type of sexual selection had prevailed, the need for physical prowess in the protection of women and children would have lessened. This, in turn, could have led to significant female participation in politics and economics, though not necessarily in a way that is imitative of male participation. In such a barely imaginable scenario, women would not have competed aggressively with one another on the basis of physical appearance and techniques of seduction; competition would have been mild in relation to what now obtains, and related more to the acquisition of accomplishments and charm than physical appearance. The female sex could then have evolved to be equal and complementary to the male sex. Instead, the impact of women's competitive strategies on their sex is directly opposite to the impact of sexual selection on men. Because groups of men could easily dominate groups of women, in certain crucial situations it would have been every woman for herself. The female sex has evolved in such a way that a woman's greatest weapon in the "struggle for existence" can be her powers of seduction. Since such strategies are individual and focused on appearance, women find that they can be defined by their sexuality alone. Female sexuality has become detached from female personhood overall, a phenomenon that is encapsulated by the "double standard" and the dichotomies revealed by the ubiquitous and unjust "Madonna-whore" cliché; it is extremely difficult, if not impossible, for a woman to be considered a "good" person as well as a sexual being. Women have to walk a social tightrope to avoid negative stereotyping, whereas men are presented with no such obstacle to full personhood.

With regard to the challenge of neo-Darwinism to theism, we have seen that there are compelling arguments for the possibility that the human species could have evolved to be a more humane and peaceful one. The crucial point is that we had a choice or choices as to how we would evolve and must, therefore, take responsibility for the result of those choices. It may be concluded, therefore, that evolutionary biology and its associated disciplines clarify the causes of evil in such a way that they lay the groundwork for empirical corroboration of theism. Even more interestingly, the evidence indicates that the fateful choice or choices that were made by our primeval ancestors strongly resonate with the notion of a primal "fall" that is common to the three Abrahamic faiths.

Some prominent evolutionary commentators have, in fact, remarked on the striking resemblance of the Darwinian and Christian accounts of human nature in particular. The main barrier to a possible synthesis of the two has been the apparently inevitable evolution of evil. This challenge has now been dealt with, along with the other main difficulty, which is the nature of the evolutionary process and the way in which it seems to rule out the occurrence of a one-off, primal sin. As I have argued, however, the decisions made by the first humans did not just affect individual morality in a once-and-for-all way; it affected the morality of the group and hence the evolutionary process itself. The Christian belief in a primal fall can now be reconceptualized as a choice or series of choices made at the very beginning of our existence as a fully human species that affected not just individual behavior but the behavioral evolution of tribes and hence the moral progress of the species itself. Within an evolutionary context, therefore, our speculations in this chapter amount to a naturalistic account of an "original sin." Now this, of course, could be pure coincidence; it may be that the Christian doctrine is just particularly insightful concerning the flaws in human nature. Had the Genesis writers known about evolution and natural selection, they would not have had to seek a supernatural explanation for human evil. Indeed, as pointed out above, some leading evolutionary commentators have asserted that the genetic explanation for evil is a good substitute for the Christian explanation. E.O. Wilson, in his foreword to *Genetics of Original Sin: The Impact of Natural Selection on the Future of Humanity* [124] makes this very point. Yet it may also be argued that the scientific evidence offers empirical corroboration of the theistic belief in a primal fall and more specifically the Christian development of it in the doctrine of an inherited "original sin." There is also the interesting fact that Wilson's own concept of hypertrophy indicates something in humanity that goes beyond the natural: the often "monstrous" results of human psyches released from the constraints of small, hunter-gatherer societies indicate human ambition on a fantastic scale. Is it feasible that purely natural descendants of the many prehuman hominid species that roamed the earth for millions of years without disturbing its rhythms could attempt a dominion over the planet that is so breathtakingly arrogant in its scope? Could a species that is totally the product of natural processes have the enormous capacity for disruption and destruction that we, the

[124] Christian de Duve with Neil Patterson, *Genetics of Original Sin: The Impact of Natural Selection on the Future of Humanity* (New Haven and London: Yale University Press, 2010).

last surviving hominid species, possess? As Yuval Noah Harari points out in his groundbreaking book *Sapiens*:

> Seventy thousand years ago, *Homo Sapiens* was still an insignificant animal minding its own business in a corner of Africa. In the following millennia it transformed itself into the master of the entire planet and the terror of the ecosystem. Today it stands on the verge becoming a god....[125]

I would argue that the level of ambition unleashed by the first humans, so massive that it could only be realized in ways that they, if only a finite species, should not have been capable of imagining, is further indication of a dimension to human nature that is beyond the material and the finite. In further support of this, there is the evidence that indicates divine intervention in the evolutionary process. The first human grouping, I would suggest, was ensouled and therefore a species with a supernatural as well as a natural destiny. With regard to my speculations above concerning the possible hubristic decisions of the first tribal leader, I would argue that such arrogance would not have occurred in a purely natural species and would have been the behavioral manifestation of a prior spiritual transgression. In the following chapter, I will attempt to synchronize the naturalistic Darwinian version of our prehistory with the more spiritual Christian account.

---

[125] Harari, *Sapiens*, p. 465.

Chapter Six
The Evolution of Adam and Eve: Homo Lapsus

The second Genesis account of creation tells the well-known story of how the first man and woman, Adam and Eve, disobeyed God by eating forbidden fruit from the tree of the knowledge of good and evil. The pair had succumbed to intense temptation from the serpent, generally considered to symbolize Satan. Theologians frequently point out that the story of Adam and Eve as portrayed in the first three chapters of Genesis does not, strictly speaking, imply that there was an "original sin" in the sense understood by Christians. Minimally, it can be read simply as a warning to the Israelites of antiquity to resist the lure of the pagan practices to which they frequently succumbed; in the ancient Near East, the serpent was associated with fertility cults. For Jews, the story is a myth that tries to explain the introduction of evil and imperfection into creation, analogous to other such archetypal myths from the surrounding cultures. Symbolically, the tree of knowledge represents a limit on humanity, a recognition that its fate is in God's hands. In his desire to be like God, Adam transgressed the boundaries of his creaturely status and attempted to be master of his own fate. The guilt and shame that the man and woman experience after their transgression result as much from their betrayal of God's trust in them as from the knowledge they have gained. The attempt by Adam and Eve to conceal their nakedness symbolizes primarily the destruction of a previously unblemished relationship between God and humanity; with the damage done to this relationship, their own relationship is tarnished. They have lost their innocence and from now on will inevitably be subject to guilt and shame and the knowledge of the evil that engenders such feelings. As a punishment for their transgression, Adam and Eve are expelled from the garden and from the intimacy that had existed between them and God, though the text makes clear that God still loves them and will always look after them. They also lose their gift of immortality and the harmonious relationship that they had both with one another and with nature. In addition, after that first sin, the world is virtually inundated by sin, beginning with the murder by Adam's son Cain by his brother, Abel. Overall, the Genesis text is concerned with showing how evil came into the world: with regard to the latter, the existence of sin appears to result somehow from the loss of innocence of the first couple. This may have been as a result of its impact on them and the way they would have subsequently behaved. It could be argued that according to Genesis, the sin was transmitted behaviorally and culturally rather than biologically or spiritually through the soul, a suggestion that corresponds to the powerful influence that culture would have had at the earliest stage of our evolution. At any rate, Judaism does not have the concept of a hereditary sin such as that found in Christianity. The idea of a hereditary sin passed on by Adam to every human individual comes from St. Paul, a Jew and a trained rabbi who converted to Christianity after his famous experience on the road to Damascus. Paul was grappling with something that would have been a very serious problem for a Jew and with which the writers of Genesis did not have to contend. This was the crucifixion of Jesus of Nazareth, whom Christianity claims was the long-awaited Messiah of the Jews and the son of God. For Jews, it is a sacrilege to say that God can suffer and die. For the first Christians, the cross was also a serious embarrassment to their claims about Jesus; crucifixion was the most shameful form of execution in the Roman Empire. In trying to find an explanation of why Jesus suffered such a fate, Paul came up with the theological concept of original sin; Adam had passed on his sin to all of humanity, and Jesus, "the second Adam," had to atone for Adam's sin. Since no human being would have the capability to undo the sin of the first man, and since humanity itself still had to atone for the sin, only someone who was both God and human could do what needed to be done.

Within Christianity, the Genesis narrative of the fall continued its development from the New Testament theology of Paul through the early centuries of Christianity and on to the late medieval period when it contributed to the wars of religion. While the Reformation was triggered by the corrupt church practice of selling indulgences, its main cause was theological disagreement over the extent of the damage to human nature caused by original sin. The preeminent theologians on each side of the conflict were Thomas Aquinas on the Catholic side and Martin Luther on the reforming side. From the vantage point of a culture that is so well informed about the destructive impact on behavior and personality of material and psychological disadvantages in life and in which, as a consequence, the whole notion of "sin" has become deeply unfashionable, the differences between the two men can seem like theological hairsplitting.

Both of them based their respective solutions to the problem of evil on the thought of the fourth century St. Augustine of Hippo, who had developed the doctrine in response to the preaching of a then popular Welsh monk known as Pelagius. The latter had developed a theory that opposed the gradually developing doctrine of original sin. Human nature, he asserted, is free to do evil or not; the human will is all-powerful, and anyone who chooses is capable of living a life of moral perfection purely by his or her own efforts. Adam did not lose any of the privileges with which he was created, nor did he transmit a tendency toward sin to the whole human race; Adam's sin affects his descendants simply as a bad example that they can imitate. By the same token, the incarnation was not for salvific purposes but to provide an example to Christians of how they should behave.

The idea that human nature is free to do evil or not seems perfectly normal to us today and would probably be the default position of most who have not been brought up in any particular religious creed. Augustine, however, whose conversion to Christianity occurred only after years of arduous struggle with his sensual predilections, was imbued with a powerful sense of the sinfulness of humanity, especially in relation to carnal desire. This strong sense that the body is a source of sin was exacerbated by his love of the philosophy of Plato, which informed all of his theological endeavors. Plato considered the spiritual world to be superior to the material one, the soul to be superior to the body, and the body to be a nuisance:

> The body provides us with innumerable distractions in the pursuit of our necessary sustenance. . . . [It] fills us with loves and desires and fears and all sorts of

> fancies and a great deal of nonsense, with the result that we literally never get an opportunity to think at all about anything. Wars and revolutions and battles are due simply and solely to the body and its desires. All wars are undertaken for the acquisition of wealth; and the reason why we have to acquire wealth is the body, because we are slaves in its service.[126]

Body versus Soul

It was, in fact, Augustine who coined the phrase "original sin" in 396;[127] he wrote about it informally in his *Confessions* in an attempt to understand what he considered to be a general human helplessness in the face of temptation. Besides Plato, he had also been influenced by the theology of Paul who, like Augustine, had struggled with the conflict between flesh and spirit. Paul identifies original sin with a theological concept known as *concupiscence*. In theological discussions of original sin, concupiscence often refers to libido as meaning sexual desire or pleasure, though it can also be used to describe excessive greed for material goods and pleasures generally. It is the theological concept of concupiscence that has caused evolutionary commentators such as Ruse and Wilson to point out the striking similarity between the Darwinian and Christian accounts of human nature.

The Pelagian crisis spurred Augustine to action, and it was his elaboration of the Genesis account of Adam and Eve's sins of pride and disobedience that shaped the Western Church's understanding of the cause of evil and that remains familiar to

---

[126] Plato, *Last Days of Socrates* (Harmondsworth: Penguin, 1969), pp. 111–112.
[127] See Paul Rigby, *Original Sin in Augustine's Confessions* (Ottawa: University of Ottawa Press, 1987), p. 3.

anyone who has been brought up as a Christian, albeit one that has lost a great deal of ground to the many other competing explanations for evil. The works of the thirteenth-century Aquinas and the sixteenth-century Luther were in agreement on the main lines of Augustine's formulation in relation to both natural and moral evil; their disagreements related to the severity of the impact of concupiscence and the mode of transmission of the sin. One cause of their differences was philosophical. Aquinas defined human nature in Aristotelian terms rather than according to the Platonist perspective of Luther, an Augustinian monk. Aquinas's espousal of Aristotle's philosophy was due to its arrival in twelfth-century academic theological circles. In comparison with Platonism and its emphasis on the primacy of spirit over matter, Aristotle's teaching seemed dangerously materialistic. Aristotle was a biologist as well as a philosopher and it was the world of external reality as opposed to the spiritual world that preoccupied him. This resulted in a philosophy that is grounded in material reality. It generated enormous controversy, especially among followers of the Augustinian tradition who were adamant that they would not accept anything that contradicted Augustine. Many of the most highly esteemed theologians of the day felt that Aristotle relied too much on reason and accorded matter too much importance in relation to the all-important spiritual realm. Indeed, Aquinas's espousal of Aristotelian philosophy gained him many enemies. In Platonism, body and soul are distinct, with the body being a mere instrument of the soul; the soul wears the body like a suit of clothes, so to speak. For Aquinas, following Aristotle, human nature is a composite of body and soul in which the spiritual and material dimensions are complementary and of equal importance. The body is good, as is sexual intercourse, since it is necessary for the propagation of the species. Indeed, Aquinas held that before the fall, sensual pleasures would have been all the keener, since the body would have been more perfect; he did not favor theories that belittle the body as a mere instrument of the soul, or as something which contaminates the soul.[128] While the body needs the soul for completeness, the soul also needs the body for the acquisition of its particular natural characteristics. This account of human nature caused virulent controversy during his lifetime, comparable to the hostility encountered by Darwin, and for similar reasons. Considering how influential Aristotle's biological classifications still are, it is not hard to see that the introduction of his thought into a Christian environment in which the spiritual world took precedence over the material one was the medieval equivalent of the nineteenth century clash between Christianity and Darwinism. It was Aristotle who came up with the famous dictum that "man is a rational animal," a definition of human nature that has been highly influential in the Western philosophical tradition and, through Aquinas, in Catholic theology. In fact, the objections of Aquinas's most deadly enemy—another medieval saint, Bonaventure—prefigured, in two very important respects, those that emerged centuries later between Huxley and Wilberforce. Firstly, there is the whole question of whether or not the human species is related to other animal species and, if so, how this relates to the dignity of the species and the *imago dei*; secondly, there are the implications for human behavior and morality of our putative animal nature. As a man

---

[128] "It is not to the detriment of the soul that it is united to a body, but for the perfection of its nature." See Aquinas, *Quaestio Disputata de anima*, 2 ad. 214. Cited in F.C. Copleston, *Aquinas* (Harmondsworth: Penguin, 1965), p. 163.

of his time, Aristotle would have considered the human individual—animal or not—to have a spiritual soul; had that not been the case, Aquinas would not have attempted to integrate his philosophy with Christian theology. Unlike Plato, however, Aristotle did not prioritize the soul over the body, considering both to be complementary. Aquinas's incorporation of Aristotle's biology into his theology must have seemed startling in the thirteenth century; his faith would, after all, have been shaped by the Genesis account of creation, in which Adam and Eve are depicted as having being directly created by God, separately from and subsequent to the creation of animals, over which they were given dominion. Yet in his writings Aquinas, like Aristotle, clearly considers the human species to be an animal one with basic similarities to other animal species, albeit with rational powers. In a passage that would not be out of place in *The Descent of Man*, Aquinas describes the central role of the survival and sexual instincts in determining human nature; he also links human nature to all organic life on Earth, something that has been proven by the fact that we share some of our DNA with both animals and plants:

> There is in man, first, a tendency towards the good of the nature he has in common with all substances; each has an appetite to preserve its own natural being [survival]. . . . Secondly, there is in man a bent towards things which accord with his nature considered more specifically, that is in terms of what . . . nature teaches *all animals*, for instance the coupling of male and female, the bringing up of the young and so forth [reproduction] . . . . Thirdly there is in man an appetite for the good of his nature as rational,

> and this is proper to him, for instance, that he should know truths about God and about living in society.[129]

Only the mention of God would be out of place in *The Descent*, and even this is linked to Aquinas's emphatic assertion that we are a social species and that this has implications for how we should behave. Like Aristotle, Aquinas held that "man is a rational animal," and remains rational (and therefore fundamentally good) even after original sin; he defines the sin negatively as the lack of a supernatural gift, lost by Adam, termed "original justice."[130] This gift would have been a kind of integrating principle between body and soul: its loss causes a "state of unrighteousness," a weakness of the will, whose primary manifestation is concupiscence, understood as an excessive and irrational desire for material goods and pleasures.[131]

Interestingly, Aquinas held that because God had bestowed a supernatural destiny on a natural creature, the gift of original justice would have lifted humanity—had it remained in a state of innocence—far beyond merely natural levels of moral goodness. Before the fall, humanity would have been capable of achieving its natural moral goals without the necessity for habitual (sanctifying) grace; after the fall however, humanity cannot achieve even its natural goals, let alone its supernatural ones, without grace. The result is that fallen humanity is at

[129] Thomas Aquinas, *Summa Theologiae* (Cambridge: Blackfriars, 1964), 1a2ae, 94, 2, pp. 81–82.

[130] A phrase coined by the eleventh-century St. Anselm of Canterbury. See Rondet, *Original Sin*, pp. 148–50.

[131] In Catholic teaching, the concupiscible appetites themselves are natural and good; concupiscence results when they escape the bounds of reason and lead to selfishness and excess.

one and the same time both morally inferior and superior to a purely natural man. Inferior because he cannot achieve unaided the deeds of goodness that the latter could have achieved; superior because with supernatural assistance he can still achieve the higher goods demanded of him as a result of his supernatural destiny. Although a "natural man" was for Aquinas a hypothetical construct, it is clear that he considered an unfallen, natural person to be basically good, which lends itself to my argument that human nature in and of itself is basically good, and that the evolution of evil was an aberration within a purely natural context.

In contrast to Aquinas, Luther defined the sin more pessimistically as a deliberate, ongoing rejection of God, and a positive bias toward concupiscence. Besides the philosophical emphases, there were clear historical and personal reasons for this difference in interpretation of Augustine's doctrine. For one thing, while Aquinas had been born into a stable Christian environment that had benefited from the eleventh century reforms of Gregory the Great, Luther's early life unfolded during the outrageously corrupt papal era of the Borgias, when the church—not for the first time—was preoccupied with political rather than spiritual power. Further, in late medieval devotional practice there was a new emphasis on the suffering and death of Jesus, which would have strongly influenced the young Luther and which certainly predisposed him toward the theological path he would later take. Finally, there was Luther's religious scrupulosity, which was at least partly grounded in the fact that he had not chosen to become an Augustinian monk for the usual reasons; his original choice of career had been the law. One fateful evening, however, while a student at the University of Erfurt, he was returning to the school after a visit to his parents when a bolt of lightning struck him to the ground. Convinced that he was going to die, Luther called on St. Anne—his father's patron saint—for help, promising to become a monk if she saved him.

Having survived the lightning bolt, Luther decided to enter the religious life at once and chose to join a strict order, the reformed congregation of the Augustinians. According to his biographer Roland Bainton, he entered against his own inclination but felt under divine constraint. At first, all seemed well; Luther surrendered himself to a tranquil life of monastic discipline. However, the saying of his first Mass precipitated another terrifying religious crisis; Luther suffered a panic attack as he began to recite the introductory prayers to the Mass, quaking with fear when he came to the words, "We offer unto thee, the living, the true, the eternal God." As he wrote afterward "At these words I was utterly stupefied and terror-stricken. I thought to myself, 'With what tongue shall I address such Majesty, seeing that all men ought to tremble in the presence of even an earthly prince?'"[132] The sensation of paralyzing terror almost caused him to flee the altar.

From a twenty-first century perspective, it may be surmised that Luther's angst was due to a combination of extreme sensitivity and the superstitious fears of the era. At any rate, with what can only be described as an excruciating scrupulosity, he subsequently began to apply himself to vigils, prayers, hours in the confessional, fasting, and mortifications of the flesh in an attempt to overcome his sense of fear and alienation from God. Yet none of these works gave him inner peace or a sense of holiness. When an opportunity presented itself to take a trip to Rome on behalf of his order, Luther seized the chance with

[132] Ibid., p. 41.

gratitude, since it would open up for him a well-established route to salvation: the gaining of an indulgence. At the time of Luther, the precise good effected by indulgences had not yet been definitively settled, but popular views attributed quasi-magical powers to them. No one denies that the church of the time encouraged extravagant views of indulgences for financial gain, fueling popular superstitions by associating the dispensing of indulgences with visitation of the relics of the saints. The dispensing of the indulgence generally depended on a stipulated financial contribution as well as a viewing of the relic. As Luther well knew, Rome was the greatest treasure-house in Christendom of such relics, and he lost no time in seeking out their spiritual benefits.

Luther's Roman endeavors did not assuage his fears and worries, however. On the contrary, he began to suspect that the spiritual claims made for indulgences were a sham. Shortly after his return to Erfurt, he was moved from the monastery there to the university town of Wittenberg, where he would remain for the rest of his life. Here he was finally rescued from his spiritual agony by the vicar of his order, Johann von Staupitz; the latter decided that Luther should study for his doctoral degree, begin preaching, and become a professor of Scripture at the university. The Bible was not fundamental to theological study during Luther's time, and he was not familiar with the Scriptures. His first lectures were on the book of the Psalms, and this necessitated a full reading of the Old Testament. At that time, the latter was considered to be primarily a Christian book that foreshadowed the coming of the Messiah. Legend has it that Luther was studying it in the tower of his Augustinian monastery when he read the phrase from Psalm 22 that would change his life forever. This was the verse that Jesus spoke on the cross shortly before his death: "My God, my God, why hast thou forsaken me?" The psalm in its entirety is an expression of the utmost desolation, desperation, alienation, abandonment, and puzzlement. In reading it, Luther was touched to the core by the realization that Jesus too had experienced extreme spiritual distress, even though he, unlike Luther and all other human beings, was sinless. The insights that flooded Luther were soothing balm to his troubled soul, though they would also provoke his rebellion. The mystery of the incarnation in particular preoccupied him:

> Show me a single mortal in the whole universe, no matter how just and saintly, to whose mind it would have ever occurred that this should be the way of salvation to believe in him who was both God and man, who died for our sins, who rose and sits at the right hand of the Father.[133]

Only faith, he argued, could accept such a mysterious road to salvation; even the most brilliant philosophical or theological logician could never have predicted what Scripture reveals—that God had to become man and die to save us. His subsequent development of a faith-driven morality reflects the most radical of his insights, standing in counterpoint to both the medieval obsession with "works", and with Aquinas's synthesis between faith and reason. Luther did not claim that human beings are evil in themselves, however; far from it. Luther would agree with Aquinas that there is plenty of potential for moral goodness and decency even apart from Christianity. Non-Christian societies, he knew, had developed very high levels of civilization

---

[133] Ibid., p. 254.

and were able to formulate and keep the equivalent of the Ten Commandments. The whole point, for Luther, is that a morality based on the life and passion of Jesus of Nazareth cannot correspond to any moral system (such as that of Aquinas) that rests on premises, propositions, conclusions, or any form of casuistry; such a morality defies reason as it is normally understood:

> When I am told that God became man, I can follow the idea, but I just do not understand what it means. For what man, if left to his natural promptings, if he were God, would humble himself to lie in the feedbox of a donkey or to hang upon a cross? God laid upon Christ the iniquities of us all . . . God can be known only through Christ, but how lay hold on Christ when his ways are likewise so incredible? The answer is not by sight but by faith which walks gaily into the darkness.[134]

It should come as no surprise, then, that in a manner strongly reminiscent of Augustine's response to Pelagius, Luther reacted promptly and angrily when Pope Leo X permitted the dispensation of a profitable new indulgence to Albert of Brandenburg, a member of the powerful Hohenzollern dynasty.[135] Enraged at what he perceived to be a clerical confidence trick, Luther posted his ninety-five theses on door of the Castle Church in Wittenberg, precipitating two hundred years of war. Luther

---

[134] Ibid., p. 224.

[135] See A. G. Dickens, *The Counter Reformation* (London: Thames and Hudson, 1968) for a discussion of the papacy of Leo X and his issuing of the indulgence.

himself was excommunicated from the church he had grown to love and condemned by the Holy Roman Emperor as "an outlaw of the Empire," a charge that put his life at imminent and ongoing risk. Ultimately, Luther's testimony lived up to the highest demands of his own faith-based morality. In his person, the radical distinction between belief and faith can clearly be seen; as a man of his time, born into a culture that was steeped in Christian lore and symbol, Luther accepted without question the tenets of Christianity. It was, perhaps, only in such a culture of undiluted Christianity that a faith could emerge that was so powerful it enabled its holder to stand up against the highest authorities of the Holy Roman Empire.

Luther's excommunication was not caused by his protest against indulgences, however, but by the theology that underpinned it. As with Augustine, this theology was grounded in his prolonged spiritual struggle and reverted to a more pessimistic view of human nature than that of Aquinas. There was also a strong Christological dimension to Luther's new perspective on human evil; if God's action in Christ can seem so incomprehensible to us, he felt, our nature must be fatally damaged by Adam's sin. Like Augustine, Luther saw original sin as a positive active revolt of humanity against God whose main manifestation is concupiscence. As we have seen, Aquinas had defined original sin as a privation or lack of the grace of original justice. Original justice would have perfected human nature, which remains basically good, if wounded. A major difference between the Thomistic and Lutheran perspectives on original sin lies in the whole notion of "human nature." For Luther, after the fall, definitions such as "man is a rational animal" are meaningless. In relation to the goodness of God as revealed in Christ, asserted Luther, humanity is so irrevocably fallen that there can be no collective, definable human nature to be restored by grace in the sight of God. In relation to God, each individual man or woman must be defined simply as a potential "hearer of God's word."

A related difference between Luther and Aquinas on original sin concerns how it is transmitted. Luther held to a spiritual transmission of the sin, Aquinas to a biological one. The former maintained that all human souls were contained in Adam's and grievously contaminated by it. For Aquinas, the sin is transmitted through the medium of sexual intercourse. Human nature is made up of body and soul and comes into existence when a child is conceived; the biological material of the parents brings the body into existence, and God infuses a rational soul into each body. The soul, however, is now transmitted without the gift of original justice, and this causes a tendency toward concupiscence.

Here I would suggest that the evolutionary canvas now available to us provides a context and perspective far broader and deeper than anything available to the theologians of the medieval period. Within this new context, the differences between both men on the subject of original sin fade into insignificance, and complementarity emerges on the salient points of both perspectives. The newly identified group dimension of original sin resolves the main difference between the two theologians, which boils down to the severity of the sin and how human nature should be defined. As we have seen, Aquinas based his moral theology on Aristotle's definition that "man is a rational animal," with the caveat that this rationality is, due to original sin, damaged to the extent that it is only through grace that it can fulfill its potential. Moral failure results mainly from weaknesses in the will that hinder rational control of the sense appetites.

Luther, on the other hand, maintains that in the light of the necessity for Christ's passion and death, such generic definitions are meaningless; sin is so woven into our nature that often we are unable even to identify it. Rather, each Christian individual can only be saved by the healing words of the gospel of Christ.

While evolutionary biology does indicate Aquinas to be correct that there is a universal, definable human nature, it also indicates that a more accurate definition of it might be as follows: Man is a dangerously predatory animal with rational powers that are frequently overwhelmed by his baser instincts; to all intents and purposes this means that, in relation to salvation, he must indeed become a hearer of the word. To define original sin in terms of a wound to human rationality, it is now clear, does not convey the full extent of the harm that has been wrought by evolution; it is only now, with the help of evolutionary science, that we can even begin to catch a glimpse of what a truly rational species might look like. As we have seen, intelligence in the male sex has been biased toward cooperation in strategies of wealth acquisition and war rather than justice, while the lack of a cohesive female contribution to moral and political discourse has surely impacted seriously on our ability to shape societies that are just and peaceful. The harm to group morality caused by natural selection amounts to scientific corroboration of Luther's assertion concerning the species; while we *can* be defined as a species in a way that has implications for morality, we are a different and morally inferior version of the species that might have been. A famous analogy of Augustine's perhaps best captures the impact of concupiscence as revealed by the scientific evidence. Augustine asks us to imagine a pair of scales with two balance pans, one of which represents good, and the other evil. When the pans are properly balanced, arguments in favor of good or evil can be measured accurately and proper conclusions drawn. What, however, if extra weights are added to the pan representing evil? While the scales still work, they are now biased toward making a decision on the side of evil. For Augustine, the biased scales represent the human will after original sin, with the weights representing concupiscence; we still have some freedom of the will, but we are biased toward making an evil decision by the excessive tendency toward concupiscence that is part of the human inheritance. Our concupiscible appetites are, of course, natural and necessary; without original sin however, they would have been in perfect harmony with our spiritual nature. The bias toward excess in our sense appetites is more than just a cause of conflicts and inequalities; it also affects the actual discernment of good and evil. For Luther, the cross reveals a kind of righteousness different to what we normally mean by human righteousness; Luther refers to it as "the alien righteousness of Christ," a pattern of behavior that can only be known and understood through revelation.

With regard to the sin's transmission, evolutionary science deftly circumvents the tortuous arguments on both sides of the theological divide by revealing how it is transmitted biologically as Aquinas claimed but largely consists of a bias toward concupiscence, which is, at least partly, in the genes. The genetic component is, I would suggest, certainly the result of a spiritual fault that influenced the decision(s) of the first humans, though what precisely the relationship is between the spiritual and the biological regarding original sin cannot be further analyzed here; it is a question for new theological investigation. I would, however, suggest that the lack of relationship with God would also have had an ongoing effect on our moral evolution through the sin of pride. At any rate, the Christian belief is that all of us are born without having sinned personally, but with a bias toward concupiscence that remains, even after the guilt of original sin is removed by baptism. The effects of the sin can only be overcome by the lifelong work of grace within the sinner. The implications for the species are that while we are still capable of evolving morally, we are disadvantaged by a biological-spiritual handicap.

Adam and Eve

It is now possible to say that the similarity of the scientific account of human origins and human nature are so similar to the Christian equivalent that it amounts to empirical evidence for an "original sin" at the outset of human history and, in addition, for important elements of the Genesis story that had been cast into doubt due to the nature of the evolutionary process. The Genesis creation stories have been traditionally treated in Christianity as a mixture of historical fact and allegory. The eighteenth century emergence of the doctrine of evolution, however, began the process of polarizing these two methods of interpretation. It has become common in Christian circles to treat the stories either as largely mythological or to reject the scientific account of human origins altogether and interpret the biblical narratives literally. So, what exactly are the implications of our exploration of the moral events of prehistory for the Genesis account of creation, temptation and fall?

Firstly, the biblical assertion that a deleterious moral event marred the very beginning of our existence as a species is supported by the scientific evidence. As to the nature of the transgression, the scientific evidence also bears out the main lines of the Genesis story. For, while the moral misdemeanors of the first humans found material form in the greed and aggression that so influenced the evolutionary process, I have suggested that, behind such behavior, must lie a spiritual transgression. In the context of the millions of years of peaceful occupation of the planet by our hominid ancestors, the rampant and infinite ambitions of our species are incompatible with those of any purely natural creature. The unlimited scope of our material wants and ambitions are indications not only of a desire to be like God, but of a distorted image of a god or gods whom we have tended to perceive as immortal manifestations of our hubristic selves. It is certainly possible, of course, that the first humans, albeit that they possessed immortal souls, could have had primitive notions of the supernatural domain and were destined to evolve religiously to a more sophisticated concept of the deity. If this is what actually did occur, the situations outlined in the previous chapter still stand, and the motivation—"the desire to lord it"—remains one of pride that eventually became rampant. It is my opinion, however, that the hubris manifested by our species from the earliest times makes it more likely to have emerged in beings who knew God and wished to be godlike but who did not have "the wisdom or vision" of their Creator.[136] If this was indeed the case, the hypothetical scenarios depicted in the previous chapter can be transposed, with some adjustments regarding the timeframe in which they occurred, to the way humanity would have evolved behaviorally outside of "the garden," after

[136] Diane Bergeant and Robert J. Karris, eds., *The Collegeville Bible Commentary* (Minnesota: The Liturgical Press, 1989), p. 44.

the initial act of disobedience had begun to impact on the evolutionary process. The attitude that had led to rebellion, combined with fully developed language skills, would have led to a swifter evolution of evil in individuals and in social structures than previously suggested. For the first humans to knowingly and deliberately disobey and rebel against God, they would have needed to complete the transitionary phase described in chapter 4. In fact, the biblical depiction of Adam and Eve in the Garden of Eden, protected, nurtured, and being prepared by God for their roles as the founders and shapers of the human race, could be said to represent that transitional phase. Having succumbed to temptation and rejected God, their behavior outside the garden would have been grounded in and motivated by a level of pride way beyond their creaturely status. The desire for power and status in the first human group would have been fueled by a spiritual malaise. Indeed, the Bible portrays a world that quickly deteriorates into serious sin, beginning with the murder of Abel by his brother, Cain, and continuing on to the flood, which is Yahweh's judgement on humanity's ever-increasing rebellion against Him.

Whichever of the two scenarios obtained, the scientific evidence supports both Scripture and tradition in relation to original sin, and the eventual moral outcomes would have been the same, though more gradual in the first instance. If, as is in my opinion highly probable, it was the second scenario that occurred, another important aspect of the Genesis story of creation, temptation and fall is borne out: original sin consisted of a deliberate, primal act of disobedience by the first humans against a God with whom they were intimately acquainted. Keeping these two possible scenarios in mind, we will now consider the implications of our discussion thus far for other aspects of the Christian solution to the problem of evil.

Although we have seen that a primal sin was possible, we also now know that we have evolved and that morality evolves. If neo-Darwinism is correct in its assessment of the impact of ESSs, from a scientific point of view morality hasn't so much evolved as struck an adaptive equilibrium. If, however, we approach the same problem from a theistic angle, it is possible to say that morality could have evolved in the purposeful, Lamarckian fashion postulated scientifically by Gould and theologically by Teilhard de Chardin. Full rationality could have been achieved early in our prehistory had culture triumphed over genes at the outset of our moral evolution. In other words, as well as having free will in relation to ourselves as individuals, we also, as a species, had the freedom to shape our own evolution. What if the first humans had chosen to put cooperation ahead of self-interest? If indeed the first humans had chosen to put the long-term interests of the group ahead of individual self-interest, human intelligence would not have been so affected by the evolution of male alliances in the pursuit of power and wealth but rather by the evolution of cooperative alliances in the service of humanity. The latter would have included alliances among a cohesive female sex and moral complementarity between the sexes rather than the domination of one by the other. The more destructive strategies associated with reciprocal altruism would not have evolved; rather the emotions and instincts associated with kin selection would have predominated, deepening and spreading beyond the family circle to the immediate community and subsequently to the wider world. Eventually, tribes would have merged into peoples, and humanity would have evolved into a peaceful, kindly, and compassionate species, without war, poverty, or unjust inequalities. Instead, *Homo sapiens* also became *Homo lapsus*. A significant theological implication of our new knowledge of the evolutionary process, however, is that the classical doctrine of original sin can be situated within a progressive evolutionary framework, with the resulting possibility of achieving a more nuanced balance between the allegorical and literal dimensions of the Genesis creation narratives.

With regard to whether our species was founded by just one man and woman—the Adam and Eve of Genesis—since any new behavior must evolve within a species, it is a possibility that modern human sensibilities did originate with just two individuals, although this would mean that the children of Adam and Eve would have had to breed either with one another or with members of the more archaic species from which they had evolved. In the rabbinic literature it is held that Adam and Eve had many children who did indeed breed with each other and that this was permitted by God. What renders a founding pair unlikely, however, is the fact that we are a social species in which individuals are always part of a group. If there is one thing that is agreed on by both sides in the evolutionary debate, it is that morality evolved and evolves as much in interactions between individuals and the group as among individuals with one another. As we have seen in the preceding chapters, our hominid ancestors evolved, over millions of years, as individuals within groups; the behavior of the first humans would already have been shaped to a significant degree by the social evolution of their ancestors. Why then, we may legitimately ask, would God make an exception for the very first humans with modern sensibilities? Indeed, the fact that the "second Adam"—Jesus—founded a church that consists of a community of Christians whose aim is to change the world for the better could be said to reflect in redemptive fashion the existence of a first Adam and a first group. The strongest indication that our species began with an original group rather than an original pair is, however, the fact that the most intractable damage to morality appears to have been done at the level of the group through the establishment of ESSs. There is, I would argue, greater moral freedom at the individual level than at the group level. One of the things we have learned from evolutionary biology about moral freedom at the origin of our species is that the free will bestowed on the first humans did not just relate to the evolution of individuals. It also related to the evolution of the sexes and hence, through their complementarity, to the evolutionary form of the species itself. The scientific evidence indicates that Adam and Eve are better envisaged as, in one sense, the first group leader and his wife. In another, they may be conceptualized as the male and female sexes respectively; in this regard the Hebrew word *Adam* literally means "the human," while the name *Eve* derives from the word for "life," as in the "giver of life." It was the interactions of the very first men and women (though perhaps especially of the first leader and his wife) that determined the evolution of "Adam" (understood as the male sex) and of "Eve" (understood as the female sex). It is, of course, perfectly possible that the whole human race can claim the first leader and his wife as progenitors, especially if the latter produced many children who married within a small tribe. Even if this is the case, however, it may not be their genes or souls that are the vehicle of transmission of original sin but the impact on the human gene pool of the moral decisions that they made.

For the sake of convenience, therefore, I will refer to the first leader and his wife as Adam and Eve.[137]

Is the idea of direct creation completely incompatible with the evolutionary process? We know that hominid bodies took millions of years to evolve and were not directly created. We also know that there are striking biological differences between the anatomy of modern humans and those of the hominid species with which they coexisted for hundreds or perhaps thousands of millennia. Interestingly, Wallace believed that, besides our higher intellectual and moral faculties, there were certain aspects of the human body that natural selection could not have achieved, including human hairlessness, the shape and capabilities of our hands and feet, and the human voice. For Wallace therefore, there also may have been a certain degree of divine intervention with regard to the human body.[138] There is certainly enough evidence for the theist to speculate that the

[137] In *Did Adam and Eve Really Exist? Who They Were and Why You Should Care* (Wheaton, IL: Crossway, 2011), C. John Collins puts forward various scenarios concerning Adam and Eve and whether there was an original pair. He favors the idea of Adam and Eve being the first king and queen of a first tribe, though he also has the interesting suggestion that there may have been other hominids of the same species outside the garden living under natural conditions. God could have brought them into the same state of being as Adam and Eve, once the first couple had made their foundational moral choice. Had they obeyed God, those others would have joined them in Paradise, where they would have been guided and led according to God's will in a manner appropriate to their creaturely status. Due to the disobedience of Adam and Eve, however, God expelled the first couple from their protected existence, though they still became the leaders of humanity. Collins cites the references in Genesis to other people living outside the garden and the fact that they are depicted as sinning in solidarity with Adam rather than by transmission of sin in support of his hypothesis. This would have affected the evolution of human nature so that the original sin became part of it.

[138] Berry, *Infinite Tropics*, pp. 201–203.

final form of the human body, along with language and our full symbolic capacities, occurred simultaneously and were achieved through divine intervention when the evolutionary process had reached a certain point. If this was indeed the case, it would lend further weight to the arguments of those such as Tattersall and Berwick and Chomsky that the first human grouping was a small one from whom we are all descended. The late pope John Paul II has latterly made a similar suggestion to that of Wallace's in his remark that with regard to the first humans, "We find ourselves in the presence of an ontological difference, an ontological leap, one could say."[139] This would imply that an earlier species became truly "human"—perhaps biologically as well as intellectually—with the infusion of a spiritual dimension, or soul. Such a momentous event, in an evolutionary context, can only be described as direct creation. If, in addition, we conceptualize Adam and Eve as the evolving sexes, we come very close indeed to the Genesis creation accounts; by infusing souls into the first humans, God directly created them and, by implication, Adam and Eve. As a matter of general philosophical and theological interest, this would imply a certain degree of dualism between matter and spirit rather than the contemporary viewpoint of some theistic evolutionary theology. Influenced by de Chardin, contemporary theologians whose focus is the relationship between evolution and creation tend to base their theology on orthogenesis, the view that evolution is purposefully directed by God. The latter theological perspective either does away completely with spirit-matter dualism by arguing that matter evolves into spirit, or, as we saw in chapter 1, treats both

[139] John Paul II, "The Pope's Message on Evolution." *Quarterly Review of Biology* 72: p. 383.

as two extremes of the same substance. The evolutionary and paleoanthropological evidence however supports Aristotelian-Thomistic dualism, which asserts the complementarity of body and soul rather than the Platonic-Cartesian view that the body is inferior to the soul.

Yet another difficulty in the reconciliation of the scientific version of origins with the theological one is the existence of "natural evils," which, as we have seen, were a factor in Darwin's loss of faith. Humans, their prehuman hominid ancestors, as well as plants and animals were and are subject to bodily decay and death; what Darwin called "the war of nature" means that at any one time animals are stalking, preying on, killing, and eating one another in a never-ending cycle. When reflecting on "the struggle for existence" in nature, Darwin consoled himself with the thought that in the animal world "the war of nature is not incessant, that no fear is felt, that death is generally prompt, and that the vigorous, the healthy, and the happy survive and multiply."[140] In addition, as we now know, the death of plants and animals is essential to the healthy functioning of the ecosystem. For humans, however, there is the added and incalculably greater suffering caused by the possession of sentience and rationality. Unlike other creatures, we anticipate—and hence fear—pain, suffering, and death, and we also suffer the sorrow of bereavement. In the classical theological context, natural evil relates to the preternatural gifts lost by Adam and Eve as a result of original sin. According to Christian tradition, Adam and Eve were placed in the beautiful garden of Eden, where they lived with dominion over, and in harmony with,

[140] Cited by Richard Dawkins in *The Greatest Show on Earth* (London: Bantam Press, 2009), p. 401.

nature. In the garden, they were free from all bodily suffering and defects, possessed the gift of immortality, and were exempt from the dangers, discomforts, fears, and sorrows to which their descendants are prey—illness, accidents, environmental threats of various kinds, the fear of death, and the loss of loved ones. In a garden, the forces of nature are kept in check and at bay by the gardener. Symbolically—and perhaps factually—such a state of existence for the first humans would mean that they were exempt from the forces that drive natural selection. Science can neither confirm nor deny the biblical texts in this instance, which remain outside the reach of empirical evidence. Evolutionary biologists and other evolutionary commentators have pointed out, however, that the ability to feel physical pain is necessary for survival and cannot be separated from the ability to experience pleasure.

On the question of natural evils, the theologian who comes closest to the Darwinian worldview is, unsurprisingly, Aquinas. The latter believed that, even if Adam and Eve had not sinned, the wolf would devour the sheep. In other words, nature would still have been predatory.[141] However, although animals fought among and preyed on one another, in their original state humans would have been the masters of creation. Due to the supernatural gift of original justice, the human body was, according to Aquinas, superior in every way to its current condition, and this included greater sensibility and hence a greater capacity to experience sensual pleasure. Aquinas does not maintain that humans were immune to pain, but due to a perfected physical prowess, perfect coordination, perfection of the intelligence, special

[141] See Thomas Aquinas, *Summa Theologica* (Indiana: Christian Classics, 1981), Vol. 1, Q. 96, a. 1. ad 2, p. 486.

knowledge imparted to Adam as the progenitor of the species by God, and the special care of divine providence, they would have been able to avoid illness, pain, accidents, and natural disasters. With regard to immortality, Aquinas perceives this gift as due to the mixed spiritual and material nature of humans; while the body is naturally corruptible the soul is immortal, and so long as humans remained free from sin, the body was indissoluble "not by reason of any intrinsic vigor of immortality, but by reason of a supernatural force given by God to the soul."[142] This is not to say that Adam and Eve would have lived forever; their lifespans would have been natural. Without the physical rupture of death, however, they would have made a transition to a superior, spiritual existence. The supernatural gift of original justice, therefore, would have strengthened us both morally and physically.

At this point, I must admit that I am in agreement with tradition on the question of the preternatural gifts; if there is a benevolent, omnipotent God who has the power to do so, why would he not bestow on us the perfection of our physical and intellectual capacities, even if such "perfection" might not be exactly as envisaged by the classical tradition? The first scenario outlined above—that we began as religious primitives—would imply a physically natural species, without preternatural gifts. Again, this may indeed have been the case, though I believe the classical view to be the more likely one. If indeed the first humans did possess preternatural gifts, it also becomes clear why the loss of such gifts would have been a necessary consequence of their choices. The resulting battles for the riches necessary to satisfy human greed would have led to an unsustainable level of

---

[142] Ibid., 1a, Q. 97, art. 1, p. 489. See also Rondet, *Original Sin*, pp. 160–184.

competition because there would have been no need for cooperation against the forces of nature. Such unrestricted competition would have produced moral monsters and led to infinite demands on the resources of a finite planet.

From the point of view of those who, like Darwin, find natural evils to be an obstacle to faith, the loss of the preternatural gifts might seem irrelevant, as God would have foreseen their loss yet still permitted the world as it is to come into existence with all of the physical sufferings and dangers that it contains.[143] In this regard, Aquinas maintains that God did not will physical evils for their own sake, but he did will the creation of a universe in which bodily death and physical suffering would be part of the natural order of events and "for the perfection of the universe."[144] The latter phrase prefigures the arguments of those who, like Wallace, recognize the need for predation and death in the maintenance of ecological balance, and acknowledge the stunning beauty of the natural world that is the result. God wished to create this particular world, and in such a context, death and decay are natural phenomena. The same cannot be said, however, about natural evils that result from human stupidity; many accidents are avoidable and are due to human carelessness, culpable ignorance, or greed. Had the first humans succeeded in evolving morally through culture in the Lamarckian sense, justice and cooperation would have prevailed over greed and aggression, and hence—even without the preternatural gifts—they would have been spared much natural as well as

[143] F. C. Copleston, *Aquinas: An Introduction to the Life and Work of the Great Medieval Thinker* (London: Penguin, 1991), pp. 152–153.
[144] Ibid.

moral evil. As Tattersall has pointed out with regard to our more ecologically destructive acts:

> Behaviors that a resilient environment could simply absorb when *Homo sapiens* was thin on the ground become hugely damaging to human populations when there are seven billion of us around. What's more, the effect cuts both ways: the more enormous and complex the human enterprise grows, the more fragile it becomes. A flood that would have been a mere inconvenience to scattered groups of hunter-gatherers becomes a human tragedy of major dimensions in the crowded landscapes of Bangladesh or the Mississippi Valley.[145]

In a species driven by rapid cultural evolution grounded in the establishment of just societies, a priority would surely have been the regulation and organization of populations and societies for the maximum physical and moral benefit of all. As for moral evil itself, this was certainly not willed by God: unlike bodily decay and death, moral evil is not an inevitable result of evolutionary processes, as are certain natural evils.

The Question of Monogenetic Origin

The mode of transmission of original sin preoccupied all of the great theologians who have grappled with the doctrine. We have focused on just three, Augustine, Aquinas, and Luther, since they are preeminent in the tradition. In the past, evolution has appeared to add yet more difficulties to the theological problems

[145] Tattersall, *Masters*, p. 230.

that have beset the doctrine in this regard, as it cast into doubt the monogenetic origin of humankind and hence the universality of the sin. It was thought up until very recently that our species originated at various different times in different regions of the world, hence ruling out descent from an original pair or an original group; this would mean that original sin, as traditionally understood in Christianity, could not have occurred. Now, however, the scientific consensus is monogenetic rather than polygenetic, with the implication that there is a universal human nature. If exclusive biological descent from an original sinning pair is not necessary to safeguard the doctrine of original sin, a universal human nature is. If every human being is infected with the sin, this is due to the fact that each and every one of us is descended, through the medium of sexual intercourse and consequent reproduction, from an original sinning pair or group. If paleontology or evolutionary biology were to claim, for example, that hominids became human in various parts of the world, this would compromise the universality of original sin. As things stand, however, scientific monogenism is the moral equivalent of theological monogenism.

With our new understanding of the role of leadership and society in the evolution of morality, we know that the moral decisions taken by the first leader would have been decisive in the moral and cultural development of our species; hence, as I have argued, we do not have to be descended from Adam himself to be the inheritors of his sin. We do, however, have to be the descendants of those who agreed with and implemented his decision, either by omission or commission. In the previous chapter, I suggested that the primal sin may have resulted from the first leader and his wife wishing to "lord it over" the rest of the tribe, an event that could have set a cultural trend and caused a general ratcheting up of greed. The other, and in my opinion, far stronger possibility is that the first leader of the human species may have appropriated for himself the role of a god, convincing his wife and people (or even vice versa) that they would henceforth have no need for God. If this were true, it would mean that Adam, Eve, and the first group were all directly implicated in the sin, and that the human race is descended from the first group through the decision of the first pair that was executed by the group overall. One way or another, there were probably several rapid genetic culls that shaped human nature into its present form toward the end of the Pleistocene, between three hundred thousand and seventy thousand years ago.[146] The salient point is that by then, the original sin would have worked its way through and formed the whole surviving population of humans, causing the development of an excessively competitive culture, and driven by the behavior of those prepared to put their own interests ahead of the good of the group. We are all the descendants of this group or population. Hence each and every one of us has inherited genetic traits and strategies that enable us to survive and reproduce but that are also responsible for a tendency toward moral evil that is prior to any personal moral decision; included in the latter tendency is responsibility for the recurrence, in every generation, of the ESSs established when human nature attained its present form. Indeed, original sin may be more about what we are capable of in certain circumstances than our general behavior.

---

[146] The Pleistocene is the vast period of time during which our species evolved. It began from the start of a glacial cycle around 2.6 million years ago and lasted up to the "the last major icecap retreat about 12 thousand years ago." See Tattersall, *Masters*, pp. 147–148.

Evolutionary Biology and Christianity: A Productive Synthesis

The possibility that the human species could have evolved to become morally superior negates one of evolution's main challenges to theistic religions such as Judaism, Christianity, and Islam. In a broader religious sense it also confirms the intuition behind many other ancient mythologies that the world would have been a better place but for a primal human mishap of one sort or another. As we have seen, for Judaism the belief is in a fall rather than an inherited sin, and this is also the case in Islam, which focuses on expulsion from Paradise and the resulting susceptibility to temptation by Satan rather than a congenital sin. Of the three, Christianity is the one with the most developed account of that mishap and the greatest similarity to the Darwinian perspective on human nature; as a result, Christianity lends itself most easily to synthesis with evolution. The scientific and theological doctrines mutually expand on and explain one another. Darwinian insights into our prehistory inform us of the actual events that underpin the mythological aspects of the Genesis story and clarify the impact on human nature of the behavioral decisions taken by the first humans. Meanwhile, revelation assuages the bad news from science with knowledge that science alone could never provide. In a Darwinian context, the doctrine of original sin is good news because it provides reassurance that the damage done to human nature by natural selection is not set in genetic stone.

At this stage in our discussion it is time to point out that the question of whether we could have evolved into a more peaceful and cooperative species has not yet been asked by scientific evolutionary commentators. Such a question properly belongs only to the realms of theology and philosophy of religion. This is because, within a secular evolutionary context, morality is relative; terms such as *better* or *worse* are only relevant to our present morality, and any "improvements" to be made would take our present condition as its starting point. While even within a secular context it is of course possible to envisage ourselves as a more peaceful, humane, and cooperative species than presently obtains, morality itself can only be considered to have an objectively good dimension if evolution is indeed a process that has been set in train by a personal, beneficent deity; similarly, an evolutionary outcome that would have been objectively preferable from a moral point of view depends on the existence of such a deity. It would be extremely dangerous to try to achieve a secular definition of objective moral good since, as Nietzsche pointed out, we are so under the grip of the will to power and the various moralities that have been generated by the evolutionary process, that moral objectivity is impossible. Only theism has the option of taking the starting point of human morality offered now by the scientific evidence in order to undertake a scientific exploration of what *could* have happened as well as what actually happened at the outset of human history. For the theist, such an exploration is revelatory, not only of the will of God for the human species, but of the meaning of divine beneficence as evidenced in what we may term the objective good. In this context, the good refers to the material goodness of creation, which God destined us to enjoy, as well as the objective moral good that God intended us to achieve. Religiously speaking, therefore, it is well worth our while to explore the moral events of prehistory. In addition, both aspects of the good must be taken into account if we are to define progress itself in an objective way. In a context of moral relativism, the potential for eugenic disasters described by philosophers such as Gray and feared by evolutionary biologists such as Gould becomes a horrendous possibility. It is on the issue of how to define human progress that religious insights are most relevant to the scientific sphere, hence the necessity to define progress in a manner that cannot be easily dismissed by science.

In the following chapter, we will bring to bear on the conclusions drawn thus far the insights of leading philosophers of science. The latter discipline, in conjunction with philosophy of religion, has a great deal more to tell us about the meaning of divine beneficence and of what constitutes authentic progress. It can also demonstrate that the scientific corroboration of the Christian doctrine of a primal sin can be extended to corroboration of the existence of a beneficent deity.

Chapter 7

Consilience: The Exhilarating God of Natural Selection

As I hope has become apparent by now, the challenge of evolutionary biology to theism in general and Christianity in particular melts away once the scientific evidence is treated on its own terms. Besides providing an empirical basis for the Judaeo-Christian belief in a primeval fall, it also supplements our knowledge of what actually happened at that remote period, and of how the moral decisions made by our earliest forebears have impacted on human behavior through evolution. Having begun from a religious perspective, however, we must now proceed to clarify and further expand on the arguments made and conclusions drawn, this time from the perspective of science. To do this, we will seek to attain a deeper understanding of the relationship between religion and science as fields of knowledge. We will begin by taking a brief look at the philosophical debate on the nature of reality, generally referred to as the realism, non-realism debate. The two main philosophies on the subject of what constitutes reality are materialism and idealism. The former denies the existence of the spiritual world by collapsing everything into the material: the latter denies that there is any mind independent reality. The classic example used to illustrate both perspectives is the philosophical thought experiment contained in the question, "If a tree falls in a forest and there is nobody around, does it make any noise?" In similar vein, the idealist philosopher George Berkeley once wondered if it is possible to say that there are trees in a park when there is nobody around to see them. Albert Einstein is said to have asked his friend the physicist Niels Bohr whether he seriously believed that the moon doesn't exist if there's nobody looking at it. Bohr replied that it is impossible to prove that it does. Idealist philosophers argue that our senses of sound, color, taste, smell and touch are secondary qualities that exist in our minds, describing assumptions that they represent an objective reality "out there" as "naïve realism." A materialist would, of course, dismiss the idealist argument and maintain that our senses do indeed map and reflect reality. If a tree falls in a forest, it makes a sound whether there is anyone there to hear it or not.

We turn now to the work of three eminent philosophers of science whose writings have made a significant contribution to the latter debate: Karl Popper, Thomas Kuhn, and Michael Ruse.

Karl Popper: Empirical Evidence and Objective Fact

The first of these, the Austrian Karl Popper was born in Vienna in 1902 and was influenced by the logical positivism of the Vienna Circle; the latter was a philosophical movement that arose in response to advances in the sciences. Logical positivism is a philosophical position that holds to the view that scientific knowledge is the only kind of factual knowledge there is because it is dependent on sense experience and is therefore empirically verifiable. All traditional metaphysical doctrines—including religious ones—must therefore be rejected as meaningless, since they cannot be empirically verified as true or false.[147] While not totally dismissive of metaphysics, Popper also considered science to be the highest form of knowledge. Over time he became convinced that only science can be depended on to yield knowledge that is objectively true.[148] The aim of science is to chart objective reality, and scientific hypotheses and data must correspond to this reality if they are to be judged true. He referred to this kind of knowledge as "knowledge without a knower." For Popper, trees really do fall in forests and make sounds even when no one is there to hear them. Whether there is such a thing as a mind-independent objective reality is disputed philosophically. Popper, however, in his assertion that science maps reality, anticipated contemporary scientism and the consequent diminishment of the status of revelation and theology in particular.

The most well-known aspect of Popper's philosophy is his assertion that a scientific hypothesis must be capable of refutation; it must lay itself open to checks and be rejected if the evidence tells against it. Further, a successful test of a hypothesis can never be final; there is always the possibility that contradictory evidence will emerge. This falsifiability is, for Popper, the most outstanding hallmark of good science: it guarantees that hypotheses must be regularly checked against experience. Popper's perspective on knowledge cuts to the heart of the overall dispute between science and religion with its implication that religion and its accompanying theological speculations are little more than flights of fantasy, myths, and fairy tales. Religion

[147] William L. Rowe, *Philosophy of Religion: An Introduction* (Belmont: Wadsworth Publishing Company, 1993), pp. 90–93.

[148] See Karl Popper, *Objective Knowledge* (Oxford: Oxford University Press, 1972), p. 109. See also Michael Ruse, *Mystery of Mysteries: Is Evolution a Social Construction?* (Cambridge: Harvard University Press), p. 15.

cannot be treated as knowledge in any meaningful way and certainly not as objective knowledge. Yet it is now possible to say that a claim of Judeo-Christian revelation has withstood the test of Popper's criteria for what constitutes empirical evidence: it is highly probable that a determinative moral event or events at the outset of our history has left a stubborn imprint for the worse on human morality. Not only that, but the evidence also indicates that the original transgression was a spiritual one, capable of being carried out only by a creature in possession of a spiritual soul. Most remarkably of all from the Popperian perspective on knowledge is the fact that Martin Luther's revelation-inspired intuition of faith anticipated the full moral impact of the evolutionary process itself two centuries before the concept was first envisaged. As a result, his theology of original sin can link creation, natural selection and Christology in a new moral paradigm. We will discuss the implications of Luther's faith-based epiphany for knowledge later in the chapter. For now, however, we return to the Popperian perspective on original sin and its capacity to further enlighten us on spiritual matters.

Inferring the Existence of a Beneficent Deity

There is far more, of course, to the doctrine of original sin than an explanation for the existence of evil; its main purpose is to provide an explanation for how evil can be reconciled with the existence of a beneficent deity. As original sin and the existence of a beneficent deity are opposite sides of the same theological coin, so to speak, does the corroboration of one doctrine by science mean that the truth of the corresponding doctrine can also be scientifically inferred? Inference is an important scientific tool in the construction of hypotheses; the doctrine of evolution itself was inferred by Darwin from his research expeditions during the *Beagle* voyage. When Darwin described the flora and fauna of the Galapagos Islands in his notebooks, he referred to the animals and birds there as aboriginal creations, meaning that God had directly created them. He was struck, however, by the fact that each of the islands had their own typical finches and tortoises, so that a particular tortoise, for example, could be identified as belonging to a particular island from the shape of its shell, while finches could be similarly identified by slight differences in the length of their beaks. Darwin began to wonder if God would have created so many slightly differing species for a group of islands so close together, and eventually made the shocking (in the context of the time) inference that they were not directly created but had evolved from a common ancestor, with the slight differences between them due to corresponding ones in the adaptive requirements of the different islands. Darwin made other, similar inferences, from the vast amount of evidence he had amassed, eventually publishing his conclusions in his *Origin of Species*. At that point, evolution was still a somewhat tentative hypothesis; nowadays the evidence for it is so great that it is accepted in science as a fact. Evolution is such a slow process that it cannot be seen occurring in nature and had to be inferred from the evidence.

One of the most famous examples of inference in science was the nineteenth century resolution by the French Urbain Le Verrier and the British John Couch Adams of the problem caused to the validity of Newton's law of gravity by irregularities in the orbit of Uranus. Each man suggested, independently of the other, that these anomalies could be explained if the gravity of an unknown planet was causing a disturbance to the orbit of Uranus in its pathway around the sun. Each man also, within the framework of Newtonian mechanics, was able to calculate the precise position of the inferred new planet. Their inferences were proven to be true when the planet Neptune was subsequently discovered in the precise position they had calculated.

So, to answer the question asked above, logically, the answer has to be in the affirmative: yes, if the scientific evidence bears out the biblical revelation of a primal fall, then we may certainly infer the corroboration of the far more important doctrine that complements it and to which it refers. The hypothesis formulated in chapter 3, out of which the empirical evidence has emerged, derives from the biblical depiction of a personal, beneficent God who is interested in each one of us and is guiding history toward its fulfilment. That hypothesis—to the effect that the evolution of serious evil could not therefore have been inevitable and must have been the result of free choices made by the first humans at the origin of our species—would be pointless were it not for the biblical revelation of an all-good, all-powerful God. It is now possible to say, therefore, that the God revealed in the Old and New Testaments is also corroborated by the empirical evidence. In addition, the more nuanced and refined anticipation of that primal misdemeanor's impact by Augustine, Aquinas, and Luther in their respective theistic moral theologies, also empirically corroborated, permits us to infer that the Christian tradition's further speculations on the both the impact of original sin, the path to salvation, and the beneficence of God are also true.

Since God is a spiritual being, this inference cannot, of course, be backed up by physical evidence of the deity as in the case of the hypothesized planet; rather, the physical evidence itself points toward the infinite. In the medieval age, Aquinas offered proofs for the existence of God, based on inferences that were drawn from the Aristotelian science he had coopted into his theological synthesis. The best known of these are based on the idea of "the unmoved mover"—nothing can move without a prior mover, just as nothing can be caused without a prior cause. There must, therefore, be a first mover or a first cause; if not, we are forced to the conclusion that there is an infinite regress, with nothing to terminate it. To me, this is an eminently reasonable argument for the existence of a deity. The theory that the universe originated in a "big bang" around fourteen billion years ago vindicates Aquinas, in my opinion, because there must have been a "cause" of such an event and, as things stand, this cause cannot be grasped in scientific terms. If a scientifically plausible cause were to be discovered, the question of regress would remain—who or what caused the cause? In Popperian terms, however, Aquinas's hypothesis is a valid one, since it is theoretically open to falsifiability should a scientific explanation ever be found. It does not, however, constitute empirical evidence for God as we now understand it, in the way that evolution does for the doctrine of original sin and its raison d'être, which is the existence of a beneficent deity.

Nor can Aquinas's proofs tell us very much about the nature of the deity, whereas the empirical evidence provided by evolutionary science gives us plenty of scope for further inference with regard to the nature of God. We are now in a position to infer a great deal about the meaning of the good, with *the good* understood as the good that God intended us to enjoy, as well as the moral good he intended us to achieve. If the science of Aquinas's day enabled him to conceptualize God as the Unmoved Mover or First Cause, the research of Darwin and his successors surely reveals a dizzying perspective on God the creator of life and on the personal God revealed by Scripture. We now know so much more about *how* God creates; that, for example, he can achieve so much with so simple and ingenious an instrument as natural selection. We know that the creation of the world was as awesome and majestic an event as the poetic language of Genesis suggests. Creation covers a time span of fourteen billion years, "time stretching across world events (cosmogenesis), leading to the origin of life (biogenesis) four billion years ago and to the development of diverse organisms (phylogenesis)."[149] Evolutionary theory also reveals much that is specific to the creation of humanity. Contrary to the fears of Darwin himself when presenting evidence to the world for the first time of our common ancestry with other animals, evolutionary theory does not diminish but confirms the significance of the creation of humanity as depicted in Genesis. We now know that it is six million to eight million years since a split occurred in the ancestral apelike primate species that led to the separate evolution of chimpanzees and human beings. This split was followed by slow, painstaking evolutionary development, experimentation, and dead-ends as various competitor prehominid and hominid species took their chances with natural selection. Species we know about, such as *Australopithecus afarensis*, *Homo erectus*, *Homo heidelbergensis*, *Homo neanderthalensis*, and others yet to be classified or discovered. What we do know with certainty is that our species, *Homo sapiens*, was the outright winner in the evolutionary stakes. As we have seen, humanity only emerged in its modern form between 125,000 and 300,000 years ago, a blink of an eye in geological time. *Homo sapiens* took billions of years to emerge, and planet Earth had to labor to its fullest capacity to produce us. In *The Descent of Man*, Darwin writes

[149] Rainer Koltermann, "Evolution, Creation, and Church Documents," *Theology Digest*, 48.2 (Summer 2001), p. 127.

beautifully and in detail of the slow and gradual process, continuing over untold millions of years, by which the characteristics of masculinity and femininity were laid down and built up in nature.[150] Looking at these facts with the eyes of faith, the importance of humanity for God's plan is highlighted beyond our wildest dreams in the care, patience, and artistry that characterized our evolution. So, for that matter, is the importance of the material world and the body; in the context of millions of years of hominid evolution, the significance of human biology for the human essence can never again be questioned or denied. This brings us to the question of the *imago dei* and the dignity of the species. From the outset of the Darwinian revolution, there have been many who find the idea of human descent from animals to be incompatible with the biblical assertion that humanity was created in the image and likeness of God. The God of natural selection, however, resonates strongly with the God revealed in Scripture; human distaste for our animal ancestry reveals more about human pride and arrogance than about the dignity of the species. It also reflects the distaste for our physicality that has characterized much of Western philosophy and, via Plato and Gnosticism, much Christian theology. As many feminist scholars have argued, this philosophical distaste for human physicality reflects another prevalent belief in Western philosophy, namely, that women are less rational than men and closer to animals by virtue of the fact that they give birth and physically nurture their young. Moreover, if, as I have argued, a hominid species became fully human through the infusion of souls by God, the special status and dignity of human beings can never be in doubt.

[150] Darwin, *Descent*, Vol. I, pp. 253–320.

With regard to our behavioral evolution, the lifestyles of the prehominid and hominid species that preceded us are presented by science simply as empirical evidence of how its members lived. From a religious perspective, of course, they assume a different significance, for if moral evil is attributed to decisions made by the earliest humans, then the basic behavioral patterns of their prehuman ancestors must be conceived of as basically good in themselves. The spiritual soul was infused into a preexisting species that would have reflected the goodness of God's creation. Before the attainment of a spiritual soul we cannot, strictly speaking, attribute free will with its accompanying capacity for moral decision-making to the intellectually archaic species from which we emerged. We can, however, consider the similarities between them and ourselves that make us recognizably a species of the genus *Homo*. For the immense timespan during which hominid hunter-gatherer lifestyles evolved, brought about the establishment of behavioral patterns that persist in our own version of human nature and anchor it firmly to the good, both moral and natural, that God intended for us. Despite the various disruptions caused by the evolution of evils that need never have existed, these goods persist strongly and link us directly back through the mists of time, past the first humans, and back to the first bipedal apes. We have already discussed the lifestyles of our ancestral and related species in some detail; all we need do here is situate them in a religious context.

For over 99 percent of the millions of years during which humans and their prehuman hominid forebears have inhabited the planet, they lived in small, wandering tribes with low population densities. This was a world without agriculture, cities, or any of the trappings of civilized existence we now take for granted. It was in such intimate and nomadic circumstances that our survival and reproductive behaviors were molded. A slight effort of the imagination can conjure up once again scenes of cooperative hunting on the savanna, gathering in woodland areas, fishing in lakes, rivers and seas; as ancestral humans learned to control fire, and as cooking skills improved and eating became more than just a means of subsistence, there would have been increasing sociability, companionship, and even, as we have conjectured, a primitive conviviality around campfires. There would also have been the adventure of traveling long distances, exploring and adapting to many different types of landscape and climate. We know that the evolution of hominids into the most intelligent and adaptive genus on the planet was driven by a far greater cooperative effort in parenting than in the ancestral ape species; pair bonding would have brought about a hitherto unknown level of intimacy between males and females and bonded family units. The fullest flowering of the basic drives to survive and reproduce has given rise, in evolved humanity, to the development of complex societies and civilizations based around the family unit, to the worlds of domesticity, hospitality, and their associated culinary arts, to the worlds of work, education, and politics. The humble cave dwelling has been transformed into many different types of habitation, from the igloo to the most opulent of palaces due to the glories of architecture, while the human desire for aesthetically pleasing surroundings has given rise to sculpture and the visual arts. Meanwhile, the primitive stone-working skills of early hominids have developed into the various crafts, trades, and technologies. Ancient courtship and mating rituals have engendered their own associated creative activity in the genres of dance and music, inspired by the love of entertainment and by love itself; evolving linguistic skills eventually led to storytelling around campfires and to our species' love of narrative in all of its forms. Deeply bred into our nature is the desire to travel and explore; hence the triumphs, both technological and of the human spirit, in mapping the breadth, depths, and heights of the earth and, latterly, the push to go beyond the physical boundaries of the planet into space. Besides being explorers, we are also travelers; for those fortunate enough there are the joys of experiencing different lands, with their rich diversity of landscapes, cuisines, and cultures. So, while we may never know for sure if God created us originally with superior physical capacities and attributes in comparison to those we now possess, a consideration of the whole history of our species thus far reveals the myriad ways in which God intended us to enjoy our bodies and the material pleasures offered by the staggeringly beautiful world he gave to us to inhabit and explore, as well as the pleasures of civilization afforded by our higher intellectual and aesthetic powers. Science, then, provides us with an evolutionary narrative grounded in empirically verifiable events that elaborate on and confirm the goodness of creation extolled in Genesis, the value of the human body, the thirst for life, exploration and relationships. In the words of the psalmist:

> When I look at your heavens, the work of your fingers, the moon and the stars that you have established; what are human beings that you are mindful of them, mortals that you care for them?
>
> Yet you have made him little lower than God, and crowned them with glory and honor. You have given them dominion over the works of your hands; you

> have put all things under their feet, all sheep and oxen, and also the beasts of the field, the birds of the air, the fish of the sea, whatever passes along the paths of the seas. (Psalm 8:4–9)

Most of all, science reveals the basic goodness of human nature itself, made good over millions of years of evolution; in relation to moral evil, this enables us to draw the important conclusion that God, in creating us, took all eventualities into account. No matter what decisions our forebears made or might have made, there is a fundamental goodness to human nature, rooted in our two basic instincts, that is prior to any free moral decision; this premoral goodness with its associated behaviors and behavioral patterns cannot be eliminated. Moreover, the process of evolution itself has its own built-in safety mechanism in the evolutionary stable strategies it imposes when aggression gets out of hand; such strategies preserve the varieties of characteristics. Although ESSs are proportioned in a manner that is morally limiting for humans, their altruistic dimension means that they are always open to the transformative power of grace; whatever choices the first humans made, the damage done would not be irreparable. Evolutionary theory has made us privy to some of God's "bigger picture" that is often given as a partial response to the problem of evil. Philosophers of religion sometimes argue that the suffering and evil that has scarred human history might not be as devastating as it seems if we had a fuller understanding of God's plan. We will return to the question of evil in the following chapter's discussion of progress; for now, we turn to another philosopher of science who defines proof in terms of meaning and coherence rather than empirical evidence.

Thomas Kuhn: Knowledge as Paradigm

Thomas Kuhn, the twentieth-century American historian and philosopher of science, defines valid scientific knowledge not so much in terms of empirical evidence *for* something, as in a new model of knowledge or *paradigm* to which the empirical evidence gives rise. For Kuhn, empirical evidence cannot reveal an objective reality "out there" that is independent of us, whether such a reality exists or not; it is preparadigmatic and, if valid and significant, will provide the context and basis for a new paradigm. The primary sense of a paradigm is "that of a work or body of work which captures the scientific imagination—which commands allegiance from a group of workers and provides tasks for them to undertake."[151]

> Aristotle's *Physica*, Ptolemy's *Almagest*, Newton's *Principia* and *Opticks*, Franklin's *Electricity*, Lavoisier's *Chemistry*, and Lyell's *Geology*—these and many other works served for a time implicitly to define the legitimated problems and methods of a research field for succeeding generations of practitioners. They were able to do so because they shared two essential characteristics. Their achievement was sufficiently unprecedented to attract an enduring group of adherents away from competing modes of scientific activity. Simultaneously, it was sufficiently open-ended to leave all sorts of problems for the redefined group of practitioners to resolve.

---

[151] Cited in Ruse, *Mystery of Mysteries*, p. 19.

> Achievements that share these two characteristics I shall henceforth refer to as "paradigms," a term that relates closely to "normal science." By choosing it, I mean to suggest that some accepted examples of actual scientific practice—examples which include law, theory, application, and instrumentation together—provide models from which spring particular coherent traditions of scientific research.[152]

The paradigm determines the rules, challenges the boundaries and, most importantly, gives rise to normal science. By normal science, Kuhn means the day-to-day scientific work routinely carried out by scientists. Crucially for Kuhn, it is science where the paradigm is presupposed and unchallenged: "No part of the aim of normal science is to call forth new sorts of phenomena; indeed, those that will not fit the box are often not seen at all. Nor do scientists normally aim to invent new theories, and they are often intolerant of those invented by others."[153]

Every so often, however, a scientific paradigm starts to come apart; puzzles and anomalies abound, and the paradigm cannot explain them. At this point, some innovative scientist puts forward a new paradigm that solves or bypasses the difficulties of the old one. The scientific community's allegiance transfers to this newcomer, and in a short while, normal science resumes again. Scientific "progress" in the Kuhnian sense is not linear, and it does not yield a truer picture of an objective reality. Rather, our knowledge of the world becomes more

---

[152] Thomas Kuhn, *The Structure of Scientific Revolutions* (Chicago: University of Chicago Press, 1962), p. 10.
[153] Ibid., p. 24.

complex and sophisticated; it is a deepening of knowledge as much as a discovery of knowledge, and it causes a change in perceptions. For Kuhn, the world itself changes with a change of paradigm. He argues that, in fact, the world exists only insofar as it is perceived through a paradigm: "Outside the laboratory everyday affairs usually continue as before. Nevertheless, paradigm changes do cause scientists to see the world of their research-engagement differently. Insofar as their only recourse to that world is through what they see and do, we may want to say that after a [scientific] revolution, scientists are responding to a different world."[154] Since paradigms structure observation, they must define reality as it is perceived.

I would add to this that it's not only scientists who perceive a different reality when there is a change of paradigm; emergent scientific paradigms have repeatedly caused significant changes, not just in perceptions of scientific reality, but in perceptions of reality in general. The self-image of Western humanity in particular has been transformed and expanded by successive scientific revolutions, beginning with Copernicus's heliocentrism and continuing through to Darwin's proofs that humanity and the rest of the animal world share a common ancestry. In three centuries we have gone from thinking of ourselves as occupying center stage in the universe to the realization that planet Earth is relatively small in the overall scheme of things. The greatest upheaval caused to human self-perception was undoubtedly that caused by Darwin's and Wallace's theory of evolution by natural selection; apart from anything else, it dealt another blow to burgeoning human self-esteem. In the aftermath of the publication of the *Origin*, songs, advertisements, caricatures,

[154] Ibid., p. 111.

and cartoons brought the salient issues into the center of public discourse and consciousness; Darwin himself was frequently caricatured as part man, part ape. The comic element of human descent from more primitive species was exploited to the full, with Nietzsche wittily remarking that "formerly one sought the feeling of the grandeur of man by pointing to his divine origin: this has now become a forbidden way, for at its portal stands the ape."[155] Prior to Darwin, science still bowed to theology, and on the question of human origins was inseparable from it. So powerful was the Darwinian paradigm, however, that it blasted apart the preceding one which had united religion and science on the question of human origins and human morality through direct creation: in the process it also decisively sundered the two disciplines as fields of knowledge. Yet, as I have argued, the empirical evidence provided by neo-Darwinism has now generated a new—or to describe it more accurately—a reconstituted theological-biological paradigm; by permitting a new understanding of direct creation within an evolutionary context and by linking morality to both nature and Christology, it reconfigures the original paradigm between science and Christianity that, ironically, Darwinism itself tore apart in the first place.

What, then, might Kuhn make of this reconstituted paradigm? As we have seen, the scientific consensus holds that evolution itself is a proven fact and that natural selection is an important evolutionary mechanism. What is not yet finally and fully proven is whether it is the most important evolutionary mechanism; the latter question, as we have also seen, is the cause of huge controversy within evolutionary biology itself due

---

[155] Friedrich Nietzsche, *Daybreak* (Cambridge: Cambridge University Press, 1997), Sec. 49, p. 32.

to its implications for human behavior. Hence the competing hypothesis of neo-Lamarckism, which has not caught the public imagination in the way that Darwinism has. In Popperian terms, the empirical evidence is on the side of Darwinism rather than Lamarckism; for Kuhn however, such evidence would have to give rise to a harmonious and cohesive paradigm. With regard to natural selection, it is acknowledged Darwin's theory brings together into one whole formerly discrete areas of biology such as embryology, systematics, morphology, paleontology, biogeography, and behavior. Within biology itself therefore, Darwinism developed with great rapidity into a new paradigm, acting as the catalyst for a new school of biological thought. Darwin himself applied natural selection to human as well as animal behavior, and the immediate and easy assimilation of this part of his theory into public consciousness is further evidence of its accuracy. The synthesis achieved in the 1950s between the theory of natural selection and Mendelian genetics, which gave rise to population genetics and neo-Darwinism, expanded and deepened the Darwinian paradigm even further by providing genetic evidence in support of Darwin's assertion that human and animal behavior have been shaped by natural selection. The human behavioral dimension of Darwinism has not yet been accepted into the humanities, however, despite its impact on philosophy and psychology through Freud, Nietzsche, Gray, and others. This is largely due, as discussed in chapter 2, to the perceived reductionism and moral pessimism of neo-Darwinism. The newly identified ability of the Darwinian paradigm, however, to extend into and expand the Christian worldview in a way that resolves misgivings about Darwinian reductionism and moral pessimism is, in Kuhnian terms, another indication of its truth and authenticity. A further indication would be that it does not represent linear progress in knowledge, but presents a broader, deeper, and more complex account of the relationship between biology and theology than the one that preceded Darwin. The scientific and theological accounts of human behavior mutually expand on and explain one another, fitting together as we saw in the previous chapter, like parts of a jigsaw puzzle. The ease and simplicity with which the synthesis between the Darwinian and Christian accounts of human nature can be achieved is a strong indication that *both* doctrines are true since it overcomes an anomaly that has dogged the Darwinian paradigm where human behavior is concerned, while at the same time dissolving the apparent obstacles presented by the evolutionary process to the Christian theory of origins. In Kuhnian terms, the compatibility of the theological doctrine with the overall Darwinian paradigm also constitutes additional evidence that natural selection is, in fact, the most significant evolutionary mechanism.

Epistemological Implications: Michael Ruse

What then are the implications for knowledge if science not only provides empirical evidence for a religious doctrine but is instrumental in generating a new theological paradigm? Within the Popperian system, science should not be able to corroborate a theological doctrine, since only scientific knowledge can be considered valid, or real; yet, as we have seen, the empirical evidence bears out the doctrine of a primal fall. The Kuhnian system would certainly be more open to the Darwinian-Christian paradigm set out previously, but even within strictly scientific parameters, Kuhn would not consider the most coherent of scientific paradigms to be representative of an objective reality "out there" that is independent of us, whether such a reality exists or not. In answer to the question, "Is science a description or report of an objective reality a real world which exists independently of human kind and which would be the same even if none of us had ever been?" the Kuhnian reply would be that since "there is no reality . . . other than that seen through and created by the paradigm, neither can we ascertain whether or not there is anything outside our perception and thought."[156]

At this point, I will call on the work of Michael Ruse, a British philosopher of science whose particular area of interest is precisely the implications of science for objectivity and subjectivity in knowledge and for an objective reality generally. Ruse has, in my opinion, made the greatest contribution thus far to the question of whether scientific knowledge can be considered to map an objective reality in a way that other forms of knowledge do not. He uses what he calls "epistemic values" to distinguish between valid scientific knowledge and what he calls pseudo-science or other forms of knowledge (such as religion) to which, in his opinion, such values do not necessarily apply. Epistemic values are defined by Ruse as "those norms or modes of reasoning that are thought to put us in touch with reality."[157]

The first of these is *simplicity* (or elegance) and this value, I would suggest, certainly applies to the Darwinian-Christian paradigm. A second epistemic value (and arguably the most important one) is *unificatory power* and its related attribute of fertility: does the hypothesis or theory bring together previously unrelated domains of knowledge under a single explanatory roof? We have already considered the unificatory power of

---

[156] Ruse, *Mystery of Mysteries*, p. 9.

[157] Ibid., pp. 32-33. Ruse draws on the work of the philosopher and historian of science Ernan McMullin in his explication of epistemic values.

Darwinism itself, which extends from biology to encompass many other scientific disciplines, and beyond science to philosophy, psychology and theology. Looking at the resulting paradigm from the theological side, the unificatory power is even more striking. The pre-paradigmatic empirical evidence corroborates not just the primal fall of theistic religion but draws together the disputed Catholic and Protestant perspectives on original sin in such a fertile manner that a whole new *theological* paradigm emerges that relates the beneficence of the deity to the material and moral goodness of humanity and creation while simultaneously increasing theological understanding of the causes and effects of evil. The latter paradigm also unifies, within a religious context, the Darwinian and Lamackian perspectives on evolution, and hence resolves the problem of Darwinian reductionism regarding our higher moral concepts and aesthetic and intellectual abilities. Perhaps even more significantly, in reuniting science and religion as fields of knowledge, it closes the epistemological gap that has been gradually increasing since the scientific revolution of the seventeenth century.

A third epistemic value is that of predictive accuracy, which includes "the power to make forecasts about what one will find in the unknown . . . [for] the theory which lets us predict suggests to us that it is not just a creation of our imagination but a reflection of something 'out there.'" If, for example, it were ever discovered that human nature did not evolve in one region and from one group or population, the paradigm would no longer be valid, since the universality of human nature would be cast into doubt and hence the classical doctrine of original sin; similarly, if punctuated equilibrium were ever proven to be the main mechanism of evolution along with its implication that human nature evolved primarily through culture (or, alternatively, if some future version of Marxism were to bring about an earthly Utopia), original sin, along with much of Darwin's speculations and the whole of the neo-Darwinist enterprise, would be refuted from the perspective of science. Ruse believes that the latter three epistemic values in particular satisfy the demand of Popper that a theory that is objectively right should be falsifiable, as well as the demand of Kuhn for coherence.

In that area of evolutionary biology where religion and science appear to collide, therefore, we can see that they synthesize into a seamless whole that greatly expands our knowledge of human nature and the relationship between humankind and God. In Kuhnian as well as Popperian terms, science provides strong corroborative evidence of the occurrence of a primal fall and the existence of a beneficent deity. It also vindicates this book's central argument, which is that the fall of the first humans is best understood as a primal moral choice, or sequence of choices, that had a damaging impact on the process of evolution. In the words of William Whewell, the great nineteenth century historian and philosopher of science—who was a lecturer and acquaintance of Darwin's during the latter's undergraduate years at Cambridge—there is a "consilience of inductions" between the theological and evolutionary doctrines. According to Whewell, "the cases in which inductions from classes of facts altogether different have . . . *jumped together*, belong only to the best established theories which the history of science contains."[158] Does this mean, however, that religion and science are equally true as fields of knowledge, equally representative of an independent reality "out there"?

---

[158] William Whewell, *The Philosophy of the Inductive Sciences* (London: Parker, 1840), pp. 2, 230.

In this particular instance it has to be said that revelation takes precedence over science with regard to human behavior; the book of Genesis preceded Darwin by thousands of years. In addition, the work of Augustine, Aquinas, and Luther preempts Darwinism by centuries. Aquinas's explanation for human good and evil anticipates Darwin and meshes harmoniously with the latter's definition of human nature; the deep conviction of both Augustine and Luther that original sin has had a serious impact on human nature is also borne out by neo-Darwinism in particular. Luther's sense that sin cannot simply be categorized as a lack of self-mastery, but is woven into and inseparable from human nature resonates with the irrational, excessive predatoriness that we now know to have evolved. All three were inspired by revelation and faith in their theologies of original sin, with Luther's intuition precipitated by revelation alone rather than by philosophical or theological speculation. Hence, in this instance, we may conclude that faith and revelation generate a higher, more objective form of knowledge than that of science.

If, as I have argued above, religious knowledge is a superior form of knowledge to science, what are the implications for the existence of an objective reality? For we cannot assert that a claim of revelation is objectively true or real just because it has been vindicated by science, for this would be to say that science has greater objectivity and provides a more reliable description of reality than does religion. We are reminded here of Pilate's question "What is truth?" and its bearing on the objectivity of scientific knowledge versus the objectivity of religious knowledge.

Popper's assertion that reality is material and that science is the only form of knowledge that can chart it objectively is called into question by science's corroboration of a theological doctrine; the fact that the empirical evidence bears out the doctrine of a primal fall would appear to indicate that reality consists of something beyond matter. I am in agreement with Michael Ruse, who holds to the Kuhnian view that scientific proofs relate to the coherence of the paradigm they generate rather than to correspondence with a reality out there. Ruse concludes that science cannot, therefore, speak directly to the philosophical debate about realism and nonrealism, which is concerned with whether or not there is an objective reality outside of human perception. This is not to deny, however, that we can speak about objectivity and subjectivity in relation to science in a way satisfactory to the Popperian no less than the Kuhnian; here Ruse relates the objectivity of science to both its rigorous methodology and its pragmatic dimension. Science has the ability to cure disease, to put men on the moon, to make atomic bombs that can kill thousands of people; it is this pragmatic dimension, as well as its methodology that distinguishes it from other forms of knowledge.[159]

What then of the coherence of the Darwinian-Christian paradigm that reaches from evolutionary science into the heart of the religious domain? I would make the suggestion at this point that the latter coherence of the physical with the spiritual *is*, in fact, an indication of an objective reality independent of our perception. The consilience of science and religion on the question of human origins and human behavior is an instance of conjunction between the material and the spiritual, in which science represents the world of material existence while at the same time enabling us to understand in greater depth the relationship between the material and the spiritual. In doing so, it also facilitates a greater understanding of reality itself. It is

---

159 Ruse, *Mystery of Mysteries*, pp. 249–255.

a fallacy to identify the material and practical with reality, as has become commonplace since the Enlightenment. To further clarify the points I have made above, we will now consider how developments in science from the seventeenth century onward have paved the way for a new understanding of reality by enabling us to distinguish between superstition and magic on the one hand and genuine spirituality on the other.

Distinguishing Between Objective Fact and Objective Reality

While certainly not the only cause of the emergence of a Western culture in which atheism has become the preferred option for many, it is my opinion that the practical achievements of science have contributed to the strong perception in Western society and culture since the Enlightenment that scientific pragmatism equates to both objectivity and reality. Yet it is only a little over three hundred years since Isaac Newton published his first masterpiece, *Mathematical Principles of Natural Philosophy*. This work, among other things, set out the law of gravity at a time when its discoverer was equally immersed in both Trinitarian theology and alchemy. *Natural philosophy* was the term applied to the discipline that is the ancestor of contemporary astronomy, biology, physics, and chemistry. In the era of the Renaissance and the Reformation, *science* simply meant knowledge and was known as natural philosophy. In the seventeenth century there was no clear demarcation between natural philosophy and theology, since the world was still considered to be God's creation. Moreover, the natural world itself was believed to contain its own mystery and magic. Apart from the superstitions of the time, which still included belief in witchcraft, spirits, and magic, Renaissance humanism—which preceded the Enlightenment—had revived interest in ancient esoteric wisdom and hermetic literature. Newton is considered to be a typical example of "that blend of fascination with a mysterious past, innovative observation and abstract thinking, which . . . began to give precedence to a new area of human enquiry over theology."[160] At that time of transition between the prescientific and the modern era, there was little distinction between the material, the magical, and the spiritual. Alchemy, for example, was a blend of magic, soteriology, Christian mysticism, astrology, and what we would now understand as chemistry. Western alchemy is considered to have contributed to the development of modern medicine as well as chemistry. Its practitioners developed a theoretical framework, terminology, and laboratory techniques that are still recognizable today; as with natural philosophy in its entirety, it is the latter, empirical dimensions of alchemy that survived the transition to Enlightenment and modernity. This was due to a crucial difference in how the men of science approached their work; while they may have been as superstitious as anyone else in the surrounding culture, their developing scientific beliefs were "tentative, not dogmatic . . . based on evidence, not on authority or intuition."[161] Most significant of all was the recognition that "the test of scientific truth is patient collection of the facts." As a result of the work of men such as Kepler, Galileo, and Newton:

> The outlook of educated men was completely transformed. At the beginning of the [seventeenth] century, Sir Thomas Browne took part in trials for

---

[160] MacCulloch, *Reformation*, p. 683.
[161] Russell, *History of Western Philosophy*, p. 514.

> witchcraft; at the end, such a thing would have been impossible. In Shakespeare's time, comets were still portents; after the publication of Newton's *Principia* in 1687, it was known that he and Halley had calculated the orbits of certain comets, and that they were as obedient as the planets had been to the law of gravitation. The reign of law had established its hold on men's imaginations, making such things as magic and sorcery incredible. In 1700 the mental outlook of educated men was completely modern; in 1600, except among a very few, it was still largely medieval.[162]

The practical dimension of science also emerged to great effect in the seventeenth century with the invention of scientific instruments such as the compound microscope, the telescope, the thermometer, the barometer, and the air pump; all of these inventions enabled scientific observation and hypothesis to become both more expansive and more precise. The greatest contribution of the seventeenth century to the banishment of superstition and magic was, however, the removal of animism from the laws of physics. In preorganized religion—and to this day in primitive tribal cultures—animism encompasses the belief that there is no separation between the spiritual and the physical or material world; spirits exist in animals, plants, rocks, geographic features such as mountains and rivers, thunder, and wind. In the pagan religions of ancient Greece and Rome, the gods inhabited rivers, mountaintops, sky and sea. In Greek philosophical thought, the planets were generally thought of as

162 Ibid., p. 522.

gods, since—although seemingly inanimate—they had motion. Aristotle himself thought that the motion of the planets indicated that they were moved by divine spirits, as any inanimate object would soon become motionless if not continuously moved by some spiritual agency. Newton's first law of motion demonstrated that lifeless matter, when once set moving, will continue forever unless there is external interference. The solar system is powered by its own momentum and its own laws, with no outside agency needed. Newton believed that the planets were originally "hurled by the hand of God," who decreed the law of gravitation. Once this had been done, however, there was no further need for divine intervention.[163]

The purging by science of magic and superstition from the material world has continued ever since, with a second major landmark for scientific utility occurring in the nineteenth century with the invention of steam, electricity, and the machine production that led to the industrial revolution. This has led to a huge increase in the sense of human power, particularly over the forces of nature; where once nature was perceived to be inhabited by capricious gods who were to be feared and placated due to the damage they could inflict, it was now increasingly considered as raw material for human manipulation. In addition, the discovery of the invisible forces, energies, and wavelengths revealed by the harnessing of electricity, the invention of the telephone and X-rays showed that the invisible does not have to be magical or spiritual; phenomena that would have seemed miraculous a century earlier could now be explained by the laws of science. The transformation wrought on human sensibilities by the amazing practical powers of science in conjunction with its banishment of

---

[163] Ibid., p. 523.

superstition have, I believe, imbued many of us in the West with a powerful sense that only the material world constitutes objective reality; the spiritual is confused with the banished superstitions of the past, and hence deemed to be nonexistent. As I have argued above, however, while science can certainly be objective in its mapping of the material world, it also, and in accordance with its own criteria, demonstrates that materiality does not constitute the whole of reality. Hence we cannot say that a claim of revelation is objectively true and indicative of an objective reality because it has been vindicated by science. The reverse is the case: the material, finite world can be shown to partake in objective reality because it points toward the infinite; it has been accorded objectivity by revelation in its relationship to spiritual reality. Scientific objectivity is in a very different category of knowledge, for example, to moral objectivity, which touches on both the material and the spiritual. Science can fully map the material world and, in clearing away any excuse for superstitious or magical beliefs, also removes much of what obscures a true image of the deity and the spiritual domain. In the most crucial area—morality—where spirit and matter intersect to greatest effect, it *can* elucidate objective human reality as opposed to objective facts about the material world but only because the data of revelation enable us to interpret its findings. Objective reality is, for humankind, both material *and* spiritual; humanity occupies both the material and the spiritual domains, and evolutionary science illuminates the distinctions as well as the relationship between the two. It also points toward a higher, spiritual domain that we can glimpse but that we simply do not have the mental equipment to fully understand. Scientific objectivity, therefore, relates to the material world, while objective reality encompasses both the material and the spiritual.

In the following and final chapter, I will argue that Darwin and Wallace are the natural successors to Augustine, Aquinas, and Luther. I will also attempt to envisage a world in which, thanks in large part to their efforts, science and religion can cooperate more productively than ever before in ascertaining the future progress of the species.

Chapter 8

Christianity, Evolution, and the Meaning of Progress

It is an interesting fact that Augustine, Aquinas, Luther, Darwin, and Wallace have far more in common with one another than with the Enlightenment thinkers whose worldview predominated in the centuries that separated the theologians from the scientists. Darwin's and Wallace's theory of evolution by natural selection marked a return to a somewhat less optimistic outlook on human nature than had obtained since the rise of science in the seventeenth century. Despite approaching the same questions from seemingly opposed perspectives, the major concern of all five men was human behavior and the implications of that behavior for the existence of God and for the nature of God. All five men were also prepared to articulate their ideas and theories in the face of virulent opposition. As well as their anthropological and religious interests therefore, they share in common the courage that distinguishes those who fearlessly pursue truth, which the renowned French philosopher Paul Ricoeur considers to be a mark of those who have testified to the truth. The fact that the first three men were ecclesiastics and the fourth and fifth secular scientists is a distinction that is, I would argue, superficial within such a context.

The new paradigm of human nature that has emerged from their combined thought will, I hope, make a further contribution to the reconciliation between both disciplines that has been gradually emerging since the publication of de Chardin's *The Phenomenon of Man*.

Ironically, the very circumstances that make this new relationship possible also make it necessary, since our era is the first to achieve a level of technological progress that can either achieve great things for humanity or destroy it. As I have argued, we have reached the point where a scientifically informed, religious input into definitions of progress will be essential. How, then, may human progress be defined in this reunited material-spiritual context? At this point, we must return to a consideration of the concept of progress that now predominates in what was once known as Christendom.

In chapter 2, we looked at various sources of evidence to assess the success or otherwise of the Enlightenment project and concluded that the evidence of science and history in particular support the Darwinian focus on instinct as the defining feature of humanity rather than the Enlightenment stress on our rationality. This initial assessment is further supported by empirical evidence for a primal moral event that had a damaging effect on the course taken by evolution, while the biological-theological paradigm outlined in the previous chapter satisfies stringent epistemological criteria, drawn from the philosophy of science, that indicate the truth of both the Darwinian and Christian perspectives on human nature. Yet it is also an undeniable fact that scientific innovation, combined with liberal democratic and social democratic political systems, have improved standards of living, health, and education to a notable degree in the Western world; the masses can now also take for granted rights and freedoms that even their relatively recent forebears could not have imagined. Both liberal democracy and Marxism originated in the optimistic intellectual world of the Enlightenment; further, the Enlightenment determination to banish irrational superstitions has also given rise to a more humane environment than that which predominated in the medieval era and perhaps even in any other major civilization:

> As the Enlightenment gained ground, it spelt the end of public wars of faith, put a stop to witch persecutions and heretic burnings, and signaled the demise of magic and astrology, the erosion of the occult, the waning of belief in the literal, physical experience of Heaven and Hell.[164]

Mental illness, which had previously been treated as sinful, was judged worthy of compassion instead, while heavy prison sentences for crimes such as minor theft were greatly reduced. In addition, the social sciences—economics, sociology, psychology, and anthropology—which emerged from the Enlightenment preoccupation with developing a secular "science of man," have contributed to general understanding of the role of culture, social class, familial circumstances, and personality in behavior. Much of what would have been considered "sin" in the past can now be attributed to circumstances that are not the fault of any particular individual and can, in fact, be remedied through secular rather than religious means. How might this progress be explained within the predominantly Darwinian perspective on human nature that conforms to the scientific evidence? There is

[164] Roy Porter, *Enlightenment* (London: Macmillan Press, 1990), p. 72.

a straightforward enough answer to this question if we consider it in terms of Wallace's speculations concerning the interaction of the soul with the body in the evolution of morality. Wallace's reply to what neo-Darwinists such as Richard Dawkins have to say about the lack of a group will or purpose in populations would be that our group and species ideals are in the soul; we can easily imagine what a just and fair society would be like, and our striving toward it is not futile, though we are going forward with the disadvantage of a significant moral handicap. The institution of the European Union after the horrors of World War II was an attempt to bring a kind of peace and harmony to Europe that could not easily be threatened. I don't believe that such a vision can be explained in genetic terms alone, though the ideals of the EU's founders would have benefited from Darwinian insights as to the difficulties involved in overcoming hypertrophied tribalistic interests, as well as the tendency latent in an enormous political entity such as the EU to develop totalitarian tendencies and (as Nietzsche would have predicted) new kinds of powerful elites. If the EU is to survive, its utopian ideals will have to be reined in somewhat and a more realistic approach to human nature taken.

All eras, of course, have their particular strengths and weaknesses, virtues and vices. Great civilizations have come and gone, and the "progress" described above may not be progress in any lasting sense; there are serious external threats to our contemporary Western civilization as well as the technological threat that may lead to catastrophe from within. Moreover, even as we acknowledge the rational achievements that were seeded in the Enlightenment, we cannot forget the terrible events of the twentieth century that stand as a permanent, tragic warning to its secular vision of a perfectible humanity. Yet with regard to human rationality, it would seem that Aristotle's definition can still stand, though with major qualifications. We are animals, indisputably, and we possess rational powers. However, the Enlightenment concept of a virtually infallible reason at species level has not withstood the test of history. The complementary perspectives of Aquinas and Luther are more in harmony with all of the evidence we have considered. The full moral and spiritual improvement of which we are capable at the level of species will, as de Chardin and many others have asserted, require supernatural assistance. This does not imply, however, a ricocheting of the pendulum from unbridled optimism to unmitigated pessimism where human nature is concerned.

In the previous chapter we focused mainly on the epistemological implications of the scientific evidence concerning human origins and behavior. To understand the new insights into human nature that are the most important contribution of evolutionary science to philosophy and theology, we must turn now to a consideration of the ontological implications.

Evolution and the Nature of Being

As previously discussed, the Darwinian revolution deepened both the epistemological and ontological problems bequeathed to philosophy by science in the seventeenth century. Epistemologically speaking, Darwinism appeared to directly challenge the Genesis creation accounts and the Christian explanation for evil, while the cumulative impact of ongoing scientific discoveries and increasing empirical evidence for the truth of evolution contributed to the development of scientism. The latter, as we have seen, is the view that scientific knowledge is the only kind of knowledge that is valid. A solution for this impasse was suggested in the previous chapter's formulation of a reconstituted theological-biological paradigm on the question of human origins that shows revelation to be the highest form of knowledge. Related to the epistemological problem is the ontological one: human beings were, up until recent history, considered to be a combination of matter and spirit, or body and soul. Evolution has spawned various new philosophical definitions of human ontology, the most well known and popular of which is the materialist view: there is no distinction between mind and matter, since human cognitive faculties can now be explained scientifically, and evolution itself has removed the need for a divine explanation of human existence.

Theologically speaking, there are three main perspectives on the implications of evolution for human ontology. There is, firstly, the view of de Chardin, grounded in his assertion that matter is evolving into spirit; this view is the direct opposite of the materialist one. For de Chardin, dualism is only apparent because everything will eventually be spirit. Similar to this view is that of theologians such as John Polkinghorne, who advocates a position known as dual aspect monism. This is the thesis that matter and mind-spirit are two extreme poles of the same "stuff," with spirit being the most evolved pole. The latter viewpoints are generally associated with the Lamarckian assertion that human nature and human morality are evolving into ever higher forms. As I argued in an earlier chapter however, this viewpoint is at odds with the scientific and historical evidence and dilutes the problem of evil to such an extent that it removes the raison d'être of Christianity.

A final theological viewpoint has already been mentioned; this is the one held by Wallace and, latterly, by John Paul II. Both men, in their respective centuries and from their respective traditions, believed that at some point in the evolutionary process, God infused spiritual and cognitive powers into a pre-existing hominid species. As we have seen, the scientific evidence thus far is most indicative of the latter view; further, while we cannot make direct inferences from epistemology to ontology, the conclusion in the previous chapter that revelation is a superior form of knowledge to empirical knowledge, if true, would constitute further evidence for mind-matter dualism and the assertion that human beings are a combination of body and soul. The most important ontological insights yielded by this position relate, however, to morality; here, knowledge and being intersect in a distinctively dualistic way. Moral decision-making results, as I argued in the previous chapter, from the interaction of the spiritual soul with the material body and the brain; we are spiritual-material creatures. Epistemologically speaking, the spiritual dimension of moral decision-making bestows objectivity on our evolved morality, which would otherwise be relative. We are now also in a position to achieve a deepened understanding of human goodness and the distinction between good and evil.

We know, for example, that there is a basic goodness to human nature so strongly embedded through millions of years of evolution that it cannot be eradicated. Along with the goodness associated with our instincts and creativity, there is the immense courage, resilience, and self-discipline of which our species is capable and that also have their roots in our evolutionary past. A major implication of the recalibration of the doctrine of original sin we have undertaken is, however, that the status quo is not acceptable. As we have seen, the evidence confirms and expands on the biblical revelation of a beneficent God who wishes us to enjoy the material world to the full; the Lamarckian perspective on evolution also illuminates for us what could have happened and what must still happen, despite the obstacles to be overcome and despite the fact that we can never be as we might have been. Original sin does not mean that we have burnt our bridges, or that the world must be a "vale of tears"; as Christians, we must do everything that we can to discern and then strive to bring about the human material and psychological flourishing and well-being that God intended us to achieve. As I have also argued, however, the scientific and technological progress that has the potential to rectify much of what we have described as natural evil is dependent on moral progress; moral and technological progress are closely intertwined and if the former is lacking or deficient, the latter may cause catastrophic harm.

We have reached a point in our discussion at which the thought of another theologian from the distant past, Irenaeus of Lyons, has become relevant. The second-century Irenaeus is often contrasted with Augustine, Aquinas, and Luther on account of his more optimistic perspective on original sin. Irenaeus has attained a new relevance in the context of evolutionary progress, since he produced a religious theology of history that demonstrates a deep consciousness of progress both in theology and in history. For Irenaeus, who took a literal view of Genesis, Adam and Eve were like children: unequal combatants in the struggle against the serpent's cunning. The revolt of humanity against God was culpable, but resembled more "the caprice of a child than the pride of a devil."[165] Like de Chardin, Irenaeus perceives Christ as the destination toward which humanity travels; his Christocentric optimism is such that, as some of his critics have remarked, sin almost becomes a necessity, a precondition of the successful accomplishment of God's

[165] Rondet, *Original Sin*, p. 40; see also pp. 37–50.

redemptive plan. In this regard, Irenaeus's theology of original sin also resembles that of de Chardin, although Irenaeus in no way disputes the free will of the first humans, or that there was an original sin of disobedience against God. Hence while it cannot be said that the evolution of evil was inevitable, I believe that Irenaeus's point about the infancy of the species being relevant to early human behavior is a valuable one that is well worth pursuing further. Irenaeus was, of course, unaware of the millions of years of hominid evolution that preceded the emergence of modern humans, and I would suggest that those millions of years constituted the real childhood of the species. It was over the course of this vast timespan that the basic goodness of human nature was set down and established in a species whose intellectual faculties were not sufficiently evolved to propel them from a state of innocence into one capable of calculated and conscious evil. The attainment of modern intellectual faculties would have marked, therefore, not so much the childhood as the adolescence of the species—a notoriously rebellious phase in individual human development. If we think of the condition of the first humans as being in certain respects analogous to that of adolescence, and keep in mind that it would take generations to learn, through hard experience, the problems that can be caused by misplaced pride and confidence, the rebellion of the first humans against God becomes something with which we can identify, though in no way justify. Yet the course of history now permits us to speculate on the God's-eye view of progress. At a certain point the species would reach mature adulthood and attain the moral benefit of a long-term perspective based on a repository of knowledge concerning acts and their consequences that the very first humans would not have had, although they had a greater degree of free will. This fact, taken together with the basic goodness of human nature, would give humanity other opportunities to obey the will of God and set about rectifying the consequences of the original disobedience. I would argue that the moral history of the species is recapitulated in individual moral development and that the attribution, within Christianity, of original sin to childhood in the past and the way that children were treated as a consequence, can now be recognized as barbaric. Jesus himself made a point of stressing the innocence and goodness of children, who belong naturally in the kingdom of God, which was the focal point of his ministry. Jesus, in fact, went so far as to admonish adults that they must "become like children" if they wish to enter the kingdom. It is only in late adolescence—when the survival and sexual instincts are fully hormonally awakened—that serious sin becomes a possibility. The tendency toward sin is, however, balanced and to a certain degree offset by the idealism that is such a wonderful characteristic of human adolescence. With regard to original sin, it would perhaps be more accurate to say that we are all born with a potential for sin that is greater than it might have been and that this potential is largely latent until adulthood.

A third important factor in relation to human moral progress is globalization. Communications technology enables us to become instantaneously aware, in a way not possible since earliest prehistory, of the damage we can inflict on one another and its terrible consequences. It also permits us to become aware of the dreadful damage that has been inflicted by advances in science on vast numbers of people:

> Laboratory science . . . provided the tools for the industrialization of war, first in the mass killing

> machines of the First World War, then in the ruthless use of science in the Second World War to defeat the enemy through more effective technology and logistics. Finally, in the blinding flash of Hiroshima, thousands of man years of scientific research and development rendered an enemy city flat in a microsecond.[166]

Whereas the first humans would have had to harden their hearts to inflict suffering on their fellow tribesmen and women—suffering that they would also have had to contemplate at first hand—as time passed, and humans spread around the globe, much human suffering could occur far away from those who inflicted it, whether or not that infliction was deliberate. Now, after the passage of 100,000 centuries or so, we, their harder-hearted, though not uncompassionate, descendants can see the results of human aggression, greed, and cruelty on a scale that would have been unimaginable to those who made the original decision to disobey and the vast majority of those who came after them. The catastrophic events of the twentieth century, along with the looming danger of human extinction, must surely give our species, now in its early adulthood, pause for thought. Humanity is at a momentous crossroads; as we face a future in which science will have the power to achieve great good or inflict terrible harm, the desire of even the most well-meaning scientists and politicians to use technology for human flourishing will not be sufficient without a clear concept of what constitutes progress. Fortunately, in light of our new perspective

---

[166] Lisa Jardine, *The Sunday Times*, April 8, 2013. Based on the BBC Radio 4 series "The Seven Ages of Science."

on the relationship between the material and the spiritual, a fuller and more authentic conception of progress is, I believe, now possible. Since I am writing from within the Christian tradition, my many unsuccessful attempts to articulate such a conception in a clear and simple way have led me to the words and stories of the master communicator himself, Jesus of Nazareth.

The Great Banquet

One of the most notable aspects of Jesus's ministry was his love of conviviality and his tendency to teach and expound his message during meals. A significant number of his parables were recounted either at table or evoked by his love of table fellowship. Moreover, one of the most radical aspects of Jesus's ministry was his insistence on dining with those who were considered "sinners" or were otherwise pushed to the margins of Jewish society. The gospels of both Matthew and Luke depict a grand dinner that Jesus attended at the house of a leading Pharisee, where he taught about hospitality generally, urging his fellow guests to respect hospitality when it is offered to them and to accept it in a spirit of gratitude and humility. He also encouraged them, when giving banquets themselves, to "invite the poor, the crippled, the lame and the blind," rather than "relatives or rich neighbors," from whom they would expect reciprocal, prestigious invitations (Luke 14:12–13). In the Jewish society of the time those who were unfortunate enough to be poor or afflicted with congenital illness were counted as sinners along with those whose occupations designated them as sinners: tax-collectors, prostitutes, and gamblers, among others. Jesus goes on to tell the story of a king who has invited many to a great banquet, only to find that they all have excuses as to why they cannot come, and have been either too inconsiderate or too careless to inform the host in good time. Angered, the king then orders that "the poor, the crippled, the blind and the lame" (Luke 14:21) be invited instead, so that his house will be full for the banquet. The setting for this parable, along with its main themes, means that it contains allegorical and symbolic connotations that go beyond its literal meaning; these meanings refer to the gracious initiative of God in inviting us into his kingdom, the inclusivity of that invitation, and the dangers of complacency among the religious establishment (in this case the Pharisees) who feel that their places at the table are assured. Jesus's story was inspired by the Old Testament prophet Isaiah's more literal view of a heavenly banquet; the latter's aim was to comfort the Jewish people during a time of impoverishment and oppression. Taken together with the overall emphasis placed by Jesus on the joys of meal-sharing, it poignantly evokes the hopes, dreams, and vulnerabilities of the human condition at their most basic and intense:

> On this mountain the Lord of hosts will make for all peoples a feast of rich food, a feast of well-aged wines, of rich food filled with marrow, of well-aged wines strained clear.
>
> And he will destroy on this mountain the shroud that is cast over all peoples, the sheet that is spread over all nations; he will swallow up death forever. Then the Lord God will wipe away the tears from all faces. (Isaiah 25:6–8)

Before delving into the significance of Isaiah's poetic vision, it must first be remarked that the powerful emphasis of the whole Judeo-Christian tradition on hospitality and table fellowship resonates with the millions of years of evolution that embedded the love of such pleasures in human nature. In addition, there is the fact that an earthly paradise, in which such delights could have been available to all, was possible and was willed by God. War, oppression, hunger, illness, and unnecessary death have, however, stalked human history as inseparable companions, wreaking the deepest physical and emotional pain on nations, peoples, and those at the poorer levels of society. We could have created a world of comfort and plenty for all with relative ease; now, in order to rectify the evolved situation, we will also have to grapple with the intertwined traits that have caused such injustices and tragedies: greed, aggression, and the love of power. While science is developing techniques that have the potential to eradicate hunger, it does not have the power to transform human morality for the better, and without moral transformation any such potential could be trumped at any time by the equal potential for destruction. It is worrisome that prominent neo-Darwinists are now suggesting that they *can* improve human morality through prosthetic and eugenic means. Such an attempt on the part of science conjures up truly terrifying scenarios that we will shortly discuss.

Let us imagine for a moment a possible world in which science has developed the technological means to provide food for all, so that no one need go hungry. The banishment of hunger from the world would, of course, be a great achievement, but would certainly not constitute a fulfillment of either Isaiah's vision, or that of Jesus. Isaiah's mouthwatering description of the fine foods and wines that the Lord will provide could only have been dreamed up by someone with a degree of culinary sophistication; Isaiah may well have been something of a connoisseur. The satisfaction of hunger would only be a first, if necessary, step in the attainment of a state in which "all peoples" had attained culinary ease and a degree of sophistication; there would also have to be an improvement in standards of civilization generally. Isaiah's vision goes even further than this, however, in that the feast on Mount Zion presupposes an end to all conflict and peaceful relations among all the peoples of the earth.

This returns us to what the biblical theologian John R. Donahue has termed "the banquet etiquette" of Jesus. As Donahue points out, the story of the great banquet (see Luke 14:15–24 par.) must be considered as part of a significant collection of parables set within chapters 14 to 16 of Luke's Gospel that are either given within the context of Jesus's table talk or inspired by his love of table fellowship.[167] They must also be understood in relation to the criticism of the scribes and Pharisees of Jesus in 15:2 that, "This fellow welcomes sinners and eats with them." All of Jesus's parables were, of course, part of his efforts to describe and bring about the eschatological kingdom of God that his ministry inaugurated; taken together with the Jewish tradition that inspired them, as well as their resonance with Jesus's historical ministry—the Last Supper in particular—those parables that involve meals and hospitality generally are perhaps the most evocative of all in illustrating what the kingdom of God will be like. The term *eschatological* in relation to the kingdom refers to the Christian belief that with the commencement of the ministry of Jesus it is "already" here, if "not yet" fully realized. The kingdom of God will develop through time, and it will be an inclusive kingdom whose people will not only enjoy the pleasures of the table, but also convivial warmth and

---

[167] John R. Donahue, S.J., *The Gospel in Parable* (Fortress Press, 1988), p. 140. See also chapters 14–16 of Luke's Gospel.

friendship. There will be no hunger or deprivation, either physical or emotional, for those who enter the kingdom. Jesus's parable of the great banquet radically expands on Isaiah's heavenly banquet by locating it during time as well as at the end of time. The world of the kingdom is and will be a warmer, kinder, more just world than that which Adam has created, and the criteria for entry into it will be grounded in radically different values. The allegorical and symbolic dimensions of Jesus's "banquet etiquette" further develop Isaiah's thought in implying that the virtues associated with hospitality, especially generosity, humility, and inclusivity, are relevant to salvation.

Interestingly Harari's *Sapiens* and *Homo Deus*, though written from an atheistic perspective, are highly relevant to both Isaiah's vision and the related Christian concept of the Kingdom of God. In *Deus* he remarks on the fact that a combination of science and politics has brought humanity within sight of defeating famine and hunger. As he also points out however, despite higher standards of living, the level of depression and discontent in the developed world is far higher than in traditional societies. Harari suggests that this may be because

> people in affluent societies suffer greatly from alienation and meaninglessness despite their prosperity. And perhaps our less well-to-do ancestors found much contentment in community, religion and a bond with nature.[168]

It is an unfortunate fact that the immense improvement in our material conditions since the Enlightenment era has brought in its wake the collapse of the family and the community. Even

---

[168] *Sapiens*, p. 425. See also *Homo Deus*, p. 1.

the individual freedoms we have gained may have a serious downside in that our increased behavioral autonomy makes it more difficult to make commitments and stick with them. We thus live "in an increasingly lonely world of unravelling communities and families."[169] In addition, as we saw in the previous chapter, there has been a collapse in religious belief due to science having solved so many of the problems that it was thought only God could solve. Increased material prosperity does not bring with it the peace, happiness and depth of relationship depicted in Isaiah's vision. Instead of increasing human well-being, it may well be decreasing it; the greater the material prosperity, the less value we seem to place on relationships with one another and with God. Despite the improvements in the human condition science has enabled, the concomitant loss of traditional community, family and religious structures has engendered a sense of meaninglessness that makes us unsure of what direction we should take as a species:

> We have advanced from canoes to galleys to steamships to space shuttles—but nobody knows where we're going. We are more powerful than ever before, but have very little idea what to do with all that power. Worse still, humans seem to be more irresponsible than ever. Self-made gods with only the laws of physics to keep us company, we are accountable to no one. We are consequently wreaking havoc on our fellow animals and on the surrounding ecosystem, seeking little more than our own comfort and amusement, yet never finding satisfaction . . . Is there anything

[169] *Sapiens*, p. 428.

> more dangerous than dissatisfied and irresponsible gods who don't know what they want?[170]

In this regard, the parable of the great banquet, emerging as it did from the table fellowship of Jesus and the Jewish biblical tradition of hospitality generally, reminds us that no matter how spectacular, brilliant, or technologically complex the future achievements of science may be, the things that are most necessary for human contentment and flourishing are simple and don't change. As I have also argued, the future progress that will be made possible by science will be dependent as never before on moral progress. The contribution of science to our knowledge of morality notwithstanding, the ethical dimension of progress cannot be left to scientists. As I have argued, science pertains to the definition of facts about the material world, while the ethical sense is forged in the interplay of the spiritual and the material; it relates to objective reality, not objective material facts.

The great twentieth-century philosopher and theologian Karl Rahner has written prophetically of the scenarios that may result from utopian views of the world and humanity that perceive the person in an empirical sense as a product to be perfected.[171] He points out that the seemingly boundless possibilities presented by, for example, genetic engineering have just as much potential to trap us in grotesque forms of slavery that earlier generations could not even have begun to imagine as to contribute to authentic physical enhancement and a more just world. He particularly fears the emergence of a world government of the future that, in collusion with the mass media, will engage in

[170] Ibid., pp. 465-466.
[171] Rahner, *Theological Investigations 6*, p. 17.

brainwashing the masses into thinking that technological and scientific achievements will bring universal happiness and fulfillment, solve all of the problems of humankind, and usher in a new age with no conflicts and no need of insight and freedom. Such a brainwashing would actually bring about the opposite result. It would usher in new forms of "barbarity, slavery, the totalitarian annihilation of personality and formation of a monochrome society."[172]

Back to the Future

The latter, apocalyptic vision of a totalitarian future is, I believe, the greatest danger now confronting our species. In this regard, however, the insights emerging from disciplines such as evolutionary psychology, paleoanthropology, archaeology, and, of course, evolutionary biology into human behavior will have an invaluable contribution to make in averting such a threat. This brings us to the secular dimension of future progress that pertains mainly to the social sciences and the secular institutions of society—the law, medicine, education, and politics. Increasing knowledge from these disciplines, especially knowledge that concerns human prehistory and the evolution of morality, can guide the organization of society in the future. We have already discussed Freud's contention that the earliest modern civilizations would have had to use levels of force never previously envisaged to attain and maintain control. It may be further argued that this force was, and remains, excessive. Systems of power are generally interested in far more than maintaining public order; the underlying dynamic is also, to a greater or

---

[172] Rahner, "The Experiment with Man," in *Theological Investigations 9* (London: Darton, Longman and Todd, 1972), pp. 205–224.

lesser extent, concerned with perpetuating the power, wealth, and influence of various elites. In this regard, one of the most useful insights emerging from the various evolutionary disciplines is the role of cultural and environmental conditions in the maintenance of peaceful societies. One of the reasons for the basic disagreement between Darwin and Wallace on the impact of natural selection on human nature was caused by the different tribal cultures to which they were exposed. Darwin's experience was certainly of cruel and savage tribal behavior; Wallace, on the other hand, had a more varied and intimate experience of tribal life because he spent many years living among tribal peoples. As we saw in chapter 2, he developed an enthusiastic admiration for some of the cultures with which he came into contact and was aware of the impact that natural environments, small groups, and freedom from oppression, exploitation, and the need to compete in the "rat race" could have on individual and group behaviors. Contemporary anthropological research confirms Wallace's observations. E. O. Wilson has pointed out that anthropological studies on surviving hunter-gatherer tribes show that population density has a marked effect on levels of tribal aggression. The contemporary !Kung San, for example, are an exceptionally peaceful people among whom violence among adults is almost unknown; in the past, however, when the environment was more crowded and they had to compete for resources with other tribes, their homicide rate per capita equaled that of Detroit and Houston. Similarly, the Semai of Malaya are, most of the time, "innocent of even the concept of violent aggression. Murder is unknown, no explicit word for kill exists . . . children are not struck, and chickens are beheaded only as a much-regretted necessity." When recruited by the British colonial government to fight against Communists in the 1950s, their men didn't know that soldiers are supposed to fight and kill; yet once out of their nonviolent society and ordered to kill, they were "swept up in a sort of insanity which they call 'blood drunkenness' that led them into an orgy of killing. As Wilson points out, while humans have a definite hereditary predisposition to aggression, the latter trait appears or disappears according to population densities so that "the most peaceable tribes of today were often the ravagers of yesteryear."[173] Similarly, Chris Stringer believes that the relative peacefulness and egalitarianism of many hunter-gatherer tribes is also due to their small numbers, since small numbers facilitate cooperative efforts by the group to resist those individuals who try to exert excessive dominance; he suggests that the optimum size of a group that can function meaningfully on such a personal and social level is between 100 and 220 people, which is about the size of the larger hunter-gatherer aggregations.[174]

While we cannot revert to being hunter-gatherers, in light of such emerging information on the conditions and milieus in which we evolved it should, surely, be possible in the future to create societal and cultural conditions that will reduce the scope for destructive hypertrophic manifestations so that our more peaceful qualities can be nurtured and inequalities reduced. Related to the latter point is E. O. Wilson's observation that, in the immediate cultural aftermath of the crackdown on aggression that would have characterized the earliest modern human attempts at civilization, there would have evolved conformist personalities; he points out that "an unthinking submission to the communal will remains among the most emotionally potent

---

[173] E. O. Wilson, *On Human Nature*, pp. 100–101.
[174] Chris Stringer, *The Origin of Our Species*, p. 113.

virtues among 'good' people in the mainstream of society." This type of group conformity is a form of tribalism and related to individual self-interest; to be accepted as part of the group over against unworthy "outsiders" increases individual fitness. Interestingly, Wilson considers that the evolution of religion itself was crucial to the emergence of conformist personalities and the "readiness to be indoctrinated" by the group will. In support of what he believes to be the biological basis of religious allegiance, Wilson makes the interesting point that such allegiance can operate even when there is no theology, as in the mass rallies and rituals that marked Communism. "The May Day rallies of T'ien An Men Square would have been instantly understood by the Mayan multitudes," while a good Bolshevik was considered to be one who "will readily cast out from his mind ideas in which he has believed for years. A true Bolshevik has submerged his personality in the collectivity, the 'Party,' to such an extent that he can make the necessary effort to break away from his own opinions and convictions and can honestly agree with the Party—that is the test of a true Bolshevik."[175] While I would disagree with Wilson on the purely biological basis of religious belief, his comments shed light on the impact of evolution on religion.

In my opinion, what is needed for future progress will not be enforced or voluntary conformity and the suppression of personality but educational systems that encourage the holistic unfolding of individual personalities and instill habits of critical thinking. It is also essential, I believe, that every effort be made to determine which destructive behaviors can be offset by favorable cultural environments, which can be changed, and which

---

[175] Wilson, *On Human Nature*, pp. 184–185.

are intractable and impervious to cultural manipulation. The type of political affirmative action to combat sexism—among other "isms" that denote intolerance—already underway in many Western societies surely offers enormous scope for cultural transformation by providing an at least partial remedy for the damage done to the complementarity of the sexes at species level, damage that cannot now be biologically undone, but for which evolution through culture can certainly go a long way toward rectifying. If the task of religion will be to spiritually inspire and empower individuals and societies in the struggle to overcome the evils caused by natural selection to our instincts and rationality, it will be the task of education and politics to find ways of drawing out and enhancing the goodness that is already there. This task may be a lot easier than genetic determinism would suggest; we now know from the up-and-coming discipline of evolutionary developmental biology (a discipline that is claimed by both sides of the intrascientific conflict) that genetics isn't solely a matter of determinism, as gene regulation is an intrinsic part of our biology. Some genes, for example, may never be turned on—or off, as the case may be—depending on the environment in which an individual organism finds itself. Further, because of neural plasticity, we are able to relearn even behaviors that are genetically hardwired.

With regard to the therapeutic role of religion in healing the type of sin that is impervious to political or cultural remedies, we turn to the Christian concept of *agape* love that will characterize it. *Agape* love is a kind of super altruism, one that is unconditionally self-sacrificial and that commands us to love even our enemies. It is a type of love that is free of and completely detached from any biological payoff. In the following quote we can see that even the love associated with kin selection, considered by Darwin to be the highest form of love and the basis of all future progress, falls short of *agape*:

> Do not think that I have come to bring peace to the earth; I have not come to bring peace, but a sword. For I have come to set a man against his father, and a daughter against her mother, and a daughter-in-law against her mother-in-law; and one's foes will be members of one's own household. Whoever loves father or mother more than me is not worthy of me; and whoever loves son or daughter more than me is not worthy of me; and whoever does not take up the cross and follow me is not worthy of me. Those who find their life will lose it, and those who lose their life for my sake will find it. (Matthew 10:34–39)

As Timothy P. Jackson points out, "Jesus set family members against each other precisely to the extent that he commands them to love all neighbors regardless of biological (genetic) or cultural relatedness. He commands his followers in other words to overcome both the selfish gene and the selfish meme."[176]

In his statement that those who find their lives will lose it and that those who lose their lives for his sake will find it, Jesus is not talking just about physical life. The authentic follower of Jesus must be prepared, as he was, to take the kind of social risks that can cause loss of one's societal status and reputation. Knowing how to be socially strategic is, of course, essential to

---

[176] Timothy P. Jackson, "The Christian Love Ethic and Evolutionary 'Cooperation'" in *Evolution, Games, and God*, Martin A. Nowak and Sarah Coakley, eds. (Cambridge, MA: Harvard University Press, 2013), p. 310.

individual "fitness" and is a key evolutionary trait for success in the world of "Adam." To take such risks flies in the face of our evolutionary trajectory as a species and is another radical mark of *agape* love. Had there been no occurrence of a primal fall, agape would not have been such a radical and risky love; we would have loved our fellow humans in a way than transcends any form of biological payoff. To rectify the moral damage that has been done, humanity will have to achieve a higher level of love than the species we might have been, a mark perhaps of necessary reparation, but also of the moral greatness of which even the "fallen" species is capable.

New Frontiers: Religion and Science Reunited

The impetus toward the reunification of religion and science, which has been gathering momentum over the past half century or so, has led us to a deeper understanding of the relationship between the material and the spiritual, a new balance between optimism and pessimism regarding human nature, a new balance between faith and reason, and an enhanced understanding of the relationship between religion and science as fields of knowledge. Most of all, it has set the stage for mutual cooperation between the two disciplines which, it is to be profoundly hoped, will usher in a new era of authentic progress for our species.

However, the relationship between religion and science will differ in one marked respect from the one that obtained previously: science will always remain an autonomous discipline in its own right due to the type of knowledge that it is; the religious beliefs or otherwise of those who engage in scientific research are irrelevant to the methodology that the scientist must employ to ascertain objective facts about the material world.

By the same token, the philosophical and theological implications of scientific research must be left to the philosophers and theologians. Finally, the compelling evidence from a variety of sources, adduced throughout this book, for the occurrence of a primal fall and the existence of a beneficent deity is no guarantee that reconciliation will occur between religion and science in the near future; nor is it certain that the schism within Christianity itself will soon be healed just because of the compatibility between the theology of Aquinas and Luther that has emerged within an evolutionary context. The greatest obstacle to reconciliation between the different groupings will be the excessive tribalism and territorialism that is as much a manifestation of original sin as is concupiscence.

At a time of unprecedented moral dilemma, however, it is imperative that these rivalries be overcome. The rewards for reconciliation will be great and will also, undoubtedly, be a catalyst for interreligious dialogue generally; for with the purging of the superstitions of the past, the stage is now set for a closer, more mature relationship between humanity and God than has been possible up to now. This renewed and deepened relationship will also, inevitably, be a more challenging one. To quote the Quaker theologian and philosopher Elton Trueblood, "Faith is not belief without proof, but trust without reservations." A more mature relationship between God and humanity is bound also to be a more demanding one.

Coda

The conclusions reached in this book are dependent on an affirmative answer to the following two questions in particular: Are we all descended from an original group or population in Africa? Did a cognitive revolution take place in our species that separated Homo sapiens from its hominid cousin species? While the empirical and related paradigmatic evidence for an affirmative answer to each question is compelling, palaeoanthropology is one of the most rapidly developing and exciting areas of evolutionary research at the moment, with new fossil discoveries expanding on previous knowledge now occurring on a regular basis. Africa itself still contains large areas that remain to be explored and excavated, and more light will continue to be shed on our origins. Now is a good time to remind ourselves that Wallace, co-discoverer of natural selection, was the first to maintain that it provides empirical evidence for the existence of a transcendent dimension to human existence. In the same way that science will never be able to disprove the God hypothesis, neither will it be able to disprove what was so clear to Wallace, namely that there had to have been divine intervention in the evolutionary process where our species is concerned. This will always give evolutionary theists the right to consider and evaluate future scientific evidence through the lens of faith. As Hans Madueme has pointed out in his discussion of the relationship between science and religion, from a faith perspective it can be argued that even if seemingly contradictory evidence emerges in the short term, conflicts on the issue will eventually be resolved in favor of the theological account of our origins. Doctrines that are foundational and crucial to Christianity cannot be held hostage to shifts in the scientific evidence.[177] Indeed we have already seen that what appeared to be an irreconcilable conflict between the Darwinian and Christian accounts of human origins has dissolved into synthesis with the contribution of a century and a half of new evidence.

Such harmonization, thanks to the endeavors of generations of theologians and philosophers, has emerged at a fortuitous time. In the rapidly evolving technological landscape of the twenty-first century there is a looming danger that, as remarked in the previous chapter, we are heading as a species toward a scenario in which the human person will be considered as no more than an economic unit or a product to be perfected. It is vital therefore that theologians and philosophers of religion have the wherewithal to advance arguments about morality and progress that cannot be dismissed out of hand by even the most skeptical scientists and that can check the growing predominance of scientism in public discourse. It is my hope that Homo Lapsus can make a useful contribution to the literature on this subject.

We turn now to the implications of our conclusions for the religious domain itself. In comparison to the medieval era,

---

[177] See Hans Madueme, "'The Most Vulnerable Part of the Whole Christian Account': Original Sin and Modern Science", in Madueme and Reeves (eds.), *Adam, the Fall and Original Sin: Theological, Biblical, and Scientific Perspectives* (Grand Rapids: Baker Academic, 2014, ebook edition), pp. 225–250.

in many respects a golden age for religion that produced an extraordinarily rich combination of spiritual endeavor and theological brilliance, we live in a world that is now largely devoid of mystery. Not only has science solved many of the problems that it was formerly thought only God could solve, but it has removed any excuse to believe in magic and superstition. In the prescientific era, magic, superstition, and spirituality were—as we saw in the case of Luther and even Newton—interchangeable in many respects. Natural philosophy, the discipline that was the precursor to contemporary science, focused on the natural world as opposed to the supernatural domain that was the concern of theology. The natural world was still considered to be God's creation, however, and had its own magical and mysterious qualities. While the banishment of belief in magic and superstition has certainly been for the best overall, in conjunction with the rise of scientism it has created a spiritually impoverished—albeit materially enriched—one-dimensional form of existence in the Western world. Christianity, the predominant religion of the West, has itself lost much of its spiritual power since the Reformation. Since that time, Roman Catholicism in particular has tended to suppress theological speculation that is outside the strictly orthodox, along with any input from the prophetic voice, or lay forms of religious organization and spirituality. The hierarchical priesthood of the church has gained a disproportionate power and influence over the church in general, to such an extent that it and the institution have become almost synonymous in the public mind. As a result, Christianity in its Roman Catholic and institutional forms has become dry and overly ritualized; hence the growing popularity of evangelical and charismatic forms of Christianity, where an exuberant faith perspective and lay participation are encouraged.

Thankfully, every cloud has a silver lining, and while science has certainly demystified the world, it has also got rid of much that obscured the authentic domain of religion. With the reunification of religion and science that has been kick-started by evolutionary biology, the stage is now set for a return of the supernatural dimension of existence. Karl Rahner has declared that, due to the rise of science, "the Christian of the future will be a mystic or he will not exist at all."[178] The rationalization of the Western consciousness by scientific achievements has, ironically, prepared the way for a deeper type of spirituality that in the past belonged only to the realm of the saints; it is possible to hope now that in time, as the reconciliation of science and religion gains ground, we will see the return of life-enhancing mystery and an enriched religious practice.

Before ending the journey we embarked on at the beginning of this book, it is important to consider the impact of our power-driven primeval moral choices on religion in its broadest sense. For just as science has already banished magic and superstition from religion, it can now help to further purify it by explaining why established religions can so easily tend toward corruption. According to evolutionary commentators such as Wilson and Stringer, in even the earliest prehistorical tribes, religious power would have been allied with political power to maintain tribal cohesion and loyalty. Religion and politics evolved in tandem with one another, probably through a rapid coevolution of genes and culture. Both Stringer and Wilson believe that religious beliefs and rituals were a unifying force in tribes and, in situations of conflict, beliefs that encouraged self-sacrifice for the sake

[178] Karl Rahner, *Theological Investigations 20* (London: Darton, Longman and Todd, 1981), p. 149.

of the group would have flourished. This can explain the perennial tendency in religions administered by a priesthood for the high priestly caste to ally itself with the ruling elite and to share its zeal for political power, wealth, and prestige. In the earliest systematic religions of the ancient world, the high priest was often also the king, emperor, or pharaoh and was considered to be an intermediary between humanity and the gods. Even a religion as ethically advanced as Judaism was not immune; by the first century A.D., the Jewish priesthood had become corrupted by its association with the colonizing powers of the Roman Empire. It was into this society, whose people were longing for the emergence of a politico-religious messiah to deliver them from oppression, that Jesus of Nazareth was born. Jesus, of course, took a radically different path to the one that his followers expected him to take; yet almost as soon as Christianity became the state religion of the Roman Empire, it began to ally itself with political power. Constantine, the emperor who established Christianity as the state religion, was, in his role of emperor, also pontifex maximus, or high priest of the state religion. By the medieval era, Christianity had become so embedded in the aristocratic systems of Europe that the popes began to compete politically with the Holy Roman Emperor for overall political control. Apart from the disillusionment caused to the mass of faithful believers by the resulting corruption—which in the medieval West led to the outbreak of radical Christianity that we know as the Reformation—when priestly and political power become so closely allied, religion becomes a force of social control and oppression rather than one of spiritual liberation. It should come as no surprise that Wallace, who rejected religion because he perceived it as just such another elitist force, could not identify the God he encountered during the course of his physical and intellectual odysseys with the God portrayed to him by the Christian culture in which he grew up. In addition, religions that espouse altruistic values (and Christianity is surely *the* religion of love) risk being targeted by those to whom the gospels refer as "wolves in sheep's clothing." Where there is an abundance of genuine altruism there also will be a corresponding abundance of predatory and parasitic individuals intent on taking advantage of it. Forewarned, however is forearmed; all of the insights that have and will emerge from evolutionary biology have a vital role to play in the purification of religion now and in the future.

Index

Page numbers followed by "n" refer to footnotes.

N

O

P

Q

R